Sensing Art in the Atmosphere

This book engages artistic interventions in the aerial elements to investigate the aesthetics and politics of atmosphere.

Sensing Art in the Atmosphere: Elemental Lures and Aerosolar Practices traces the potential of artistic, community-driven experiments to amplify our sensing of atmosphere, marrying attentions to atmospheric affect with visceral awareness of the materials, institutions and processes hovering in the air. Drawing on six years of research with artistic and activist projects *Museo Aero Solar* and *Aerocene*, initiated by artist Tomás Saraceno, each chapter develops creative relations to atmosphere from the studio to stratospheric currents. Through narrative-led writing, the voices of artists and collaborators are situated and central. In dialogue with these aerographic stories and sites, the book develops a notion of *elemental lures*: the sensual and imaginative propositions of aerial, atmospheric and meteorological phenomena. The promise of *elemental lures*, Engelmann suggests, is to reconcile our sensing of atmosphere with the myriad social, cultural and political forces suspended in it. Through tales of floating journeys, shared envelopes of breath and surreal levitations, the book foregrounds the role of art in crafting alternative modes of perceiving, moving and imagining (in) the air.

The book ends with a call for elemental experiments in the geohumanities. It makes an important and original contribution to elemental geographies, the geohumanities and interdisciplinary scholarship on air and atmosphere.

Sasha Engelmann is a creative geographer engaged in environmental sensing through artistic, collective and collaborative practices. She is Lecturer in GeoHumanities at Royal Holloway University and a long-term member of the *Aerocene Community*. With Sophie Dyer she leads the feminist amateur radio project *Open Weather*.

T0203635

Ambiances, Atmospheres and Sensory Experiences of Spaces

Series Editors:
Rainer Kazig, *CNRS Research Laboratory Ambiances – Architectures – Urbanités, Grenoble, France*
Damien Masson, *Université de Cergy-Pontoise, France*
Paul Simpson, *Plymouth University, UK*

Research on ambiances and atmospheres has grown significantly in recent years in a range of disciplines, including Francophone architecture and urban studies, German research related to philosophy and aesthetics, and a growing range of Anglophone research on affective atmospheres within human geography and sociology.

This series offers a forum for research that engages with questions around ambiances and atmospheres in exploring their significances in understanding social life. Each book in the series advances some combination of theoretical understandings, practical knowledges and methodological approaches. More specifically, a range of key questions which contributions to the series seek to address includes:

*   In what ways do ambiances and atmospheres play a part in the unfolding of social life in a variety of settings?
*   What kinds of ethical, aesthetic, and political possibilities might be opened up and cultivated through a focus on atmospheres/ambiances?
*   How do actors such as planners, architects, managers, commercial interests and public authorities actively engage with ambiances and atmospheres or seek to shape them? How might these ambiances and atmospheres be reshaped towards critical ends?
*   What original forms of representations can be found today to (re)present the sensory, the atmospheric, the experiential? What sort of writing, modes of expression, or vocabulary is required? What research methodologies and practices might we employ in engaging with ambiances and atmospheres?

Rethinking Darkness
Cultures, Histories, Practices
Edited by Nick Dunn and Tim Edensor

Sensing Art in the Atmosphere
Elemental Lures and Aerosolar Practices
Sasha Engelmann

For more information about this series, please visit: www.routledge.com/ Ambiances-Atmospheres-and-Sensory-Experiences-of-Spaces/book-series/ AMB

Sensing Art in the Atmosphere

Elemental Lures and Aerosolar Practices

Sasha Engelmann

Routledge
Taylor & Francis Group

LONDON AND NEW YORK

First published 2021
by Routledge
2 Park Square, Milton Park, Abingdon, Oxon OX14 4RN

and by Routledge
605 Third Avenue, New York, NY 10017

First issued in paperback 2022

Routledge is an imprint of the Taylor & Francis Group, an informa business

British Library Cataloguing-in-Publication Data
A catalogue record for this book is available from the British
Library

Library of Congress Cataloging-in-Publication Data
A catalog record has been requested for this book

ISBN 13: 978-0-367-61491-1 (pbk)
ISBN 13: 978-0-367-22035-8 (hbk)
ISBN 13: 978-0-429-27043-7 (ebk)

DOI: 10.4324/9780429270437

Typeset in Times New Roman
by codeMantra

Contents

Figures

Acknowledgements

This book is the product of a multitude of lures. My childhood summers near the Adriatic Sea were spent feeling the "pull" of mythic winds and hot, humid weather. The spatial experiments of ecological artists and hypertext novelists led to the discipline of Geography. The aerial installation *On Space-Time Foam* drew me into my first encounter with Studio Tomás Saraceno. A full scholarship from the Marshall Aid Commemoration Commission and a Clarendon Doctoral Fellowship from the University of Oxford were powerful lures, too. Yet the most enduring lures have been those of friends, colleagues and mentors who have tethered, untethered and animated my thinking over the past ten years.

Derek McCormack taught me about atmospheres, atmospheric things, and the alluring stories that link them together. I am deeply grateful for the time we have shared writing about the elements, and sometimes floating in them, too. Through the generosity and friendship of Tomás Saraceno, I was able to spend three formative years apprenticing and collaborating in his studio in Berlin. I am happy we continue to conspire toward aerosolar futures, mobilities and politics. Harriet Hawkins has been my mentor from the earliest days of my postgraduate studies, and I am now honoured to be her colleague at the Centre for the GeoHumanities, Royal Holloway University of London. I am grateful to Nick Shapiro for enlarging so many dimensions of elemental and community worlds. Bronislaw Szerszynski has been my co-author and co-presenter and hosted me for a memorable weekend in the deep forests of Mostówka, Poland. Jol Thoms has enlivened this research, participated in every cosmic experiment and has also become my partner in life. I have benefited immensely from curatorial adventures and discussions with Sofia Lemos, Anna Sophie-Springer and Etienne Turpin. Conversations with Tim Ingold and Jamie Lorimer, my DPhil examiners, gave me confidence to pursue the publication of this book.

This book is a testament to the imaginative investment and tireless labour of practitioners of the *aerosolar arts*. I am forever grateful to my dear friends, colleagues and co-inventors of *Aerocene*, especially Bill McKenna, Camilla Berggren, Alex Bouchner, Barbara Bulc, Ewen Chardronnet, Joaquin Ezcurra, Gwilym Faulkner, Sam Hertz, Thomas Krahn, Adrian

Krell, Yelta Kom, Alice Lamperti, Sofia Lemos, Roxanne Mackie, Martyna Marciniak, Derek McCormack, Igor Mikloušić, Grace Pappas, Tomás Saraceno, Daniel Schulz, Nick Shapiro, Kotryna Slapsinskaité, Sven Steudte, Pablo Suarez, Débora Swistun and Erik Vogler. Special thanks to fellow *Aerocene* adventurer Maxi Laina for teaching me some Spanish in the middle of a salty lake at noon.

In myriad ways, the research that sustains this book was informed and made possible by members of Studio Tomás Saraceno, especially: Lars Behrendt, Fabiola Bierhoff, Ally Bisshop, Sascha Boldt, Saverio Cantoni, Tato Chaves, Connie Chester, Carola Dietrich, Sara Ferrer, Luca Girardini, Canice Grant, Martin Heller, Dario Jacopo Laganà, Sarah Kisner, Sarah Martinus, Claudia Melendez, Pepe Menéndez-Conde, Roland Muehlethaler, Martina Pelacchi, Sebastian Steinboeck, Ayşegül Seyhan, Ilka Tödt, Desirée Valdes and Zaida Violan. This includes many who have left the studio since the research began, including Stefano Arrighi, Cara Cotner, Camilo Brau, Danja Burchard, Joshua Depaiva, Edgar Diaz, Dorota Gaweda, Julia Hajnal, Adrian Krell, Tobias Lange, Veronica Lugaro, Ignas Petronis, Karina Pragnell, Javier Rosenberg, Daniel Schulz, Irin Siriwattanagul, Vasily Sitnikov, Kotryna Slapsinskaité and Emek Ulusay. I am especially grateful for the opportunity to publish a large number of colour images produced by Studio Saraceno; these images operate in close creative dialogue with the text.

In these pages, there are stories of experiments in art and architecture pedagogy at the Institut für Architekturbezogene Kunst at the Technical University of Braunschweig, upon the kind invitation of Tomás Saraceno. In this context, I am particularly thankful to Ivana Franke and Natalija Miodragovic for many insights on collaborative art practice, feminism and care. My dearest thanks to Alex Bouchner, Phillip Dreyer, Matthias Pelli and Tomi Soletic for memorable nights and mornings. I am also indebted to numerous students, especially those of the *Becoming Solar, Becoming Aerosolar* and *Becoming Pilot* postgraduate seminars, who contributed to symposia, exhibitions and projects in Berlin, Vienna, Paris and Turkey. Finally, I am grateful for dialogues with Alan Prohm, Ilka Raupach, Sina Heffner, Jessica Höfinger, Bernd Schulz and Michael Zwingmann.

Many have offered comments on chapters in this book. In particular, this book has benefited from the attention of Pete Adey, Camilla Berggren, Sophie Dyer, Joaquin Ezcurra, Harriet Hawkins, Till Hergenhahn, Rainier Kazig, Sarah Kisner, Damien Masson, Tomás Saraceno, Paul Simpson, Jatun Risba, Débora Swistun and Jol Thoms. My mom, Diana Engelmann, has contributed to the poetics of elemental lures and to the editing of all chapters in the book. My research and writing would not have been possible without the early and unique insight on atmospheres, climates and artistic interventions from my former colleagues at the arts commissioning organisation Invisible Dust: Alice Sharpe, Bianca Manu and Anne Osherson.

Many colleagues at Royal Holloway University and the Centre for the GeoHumanities have generously sounded out my ideas through conversations, Landscape Surgery seminars and symposia. They are Phil Crang, Veronica Della Dora, Felix Driver, David Gilbert, Harriet Hawkins, Anna Jackman, Innes Keighren, Oli Mould, Sofie Narbed, Cecilie Sachs-Olsen, Rachael Squire, Varyl Thorndycraft and Katie Willis. The students of my advanced level undergraduate course *Atmospheres: Nature, Culture, Politics* have contributed immensely to this book through their brilliant questions, energies and ideas, and their willing participation in the yearly launch of two *Aerocene* sculptures on the university campus.

Finally, I am inexpressibly fortunate to come from a Slavic-German family of scientists, activists and poets: thank you to my mom, Diana, my dad, Steve and my brother, Elliott.

Preface

Figure 0.1 Clouds from Zaraće bay, Hvar island, Croatia.
Source: Photography by Diana Engelmann (*née* Šatlan).

In the summertime, I often travel to a small bay called Zaraće on the island of Hvar in the Adriatic Sea, where my maternal family originates. The winds of the Adriatic are well known to local fishermen. In late August, a mythic, unpredictable wind called *Bura* ("storm" in Croatian) flows powerfully from the North over the Croatian mainland. When it reaches a range of tall mountains, thwarting its path to the sea, *Bura* travels up the steep rocky slopes. As water in the wind condenses at higher, colder altitudes, epic towers of cumulus and cumulonimbus clouds form. The clouds linger over the mountaintops like the billowing dancers of the Dalmatian folk tradition. They may seem lofty, but rarely do they move across the sea toward the island. One can watch these clouds, the materialisation of *Bura*, from a spot best reached by swimming out into the middle of the bay.

> *I pause in the middle, fill my lungs, and let my body float on the surface of the water. In quiet suspension between water and air, I observe the clouds, surrendering to their movements. Wisps move from one to the other, and bulbous forms push outward. They twirl in several different directions. After making their signs, they drift further south, and lose their distinct shapes, joining the winds again.*

These clouds have other meanings, other shapes. I was an asthmatic child living in Los Angeles when the Wars in the Balkans broke out in the 1990s. With family spread out in various parts of former Yugoslavia, my parents searched for every scrap of information on local militancy, political movements, adequate bomb shelters and the looming certainty of a US-led airstrike on Belgrade where my grandparents lived at the time. The information gathered from local news stations and relatives was painful and fragmented; it contrasted starkly with what appeared in US media. These discrepancies contributed to my mother's sense of helplessness and profound sadness, the echoes of long-distance phone calls at night.

I was too young to follow the complexities of the regional conflicts, violent land claims, and the genocide that occurred in the Balkans in the mid-1990s. These events became clearer to me in their aftermath, felt in the impressions they made on family and friends, and in the marks left on communities and landscapes that I encountered years later. However, I vividly remember when NATO forces led by the USA began bombing Belgrade in March 1999 during *Operation Noble Anvil*, known ironically in Yugoslavia as *Milosrdni anđeo* (*Merciful Angel*). I remember stories from relatives about how many bridges, churches and hospitals had been hit. I learned two decades later that *Operation Noble Anvil* was a case of "risk transfer" from military forces to civilian populations due to the use of satellites for remote weapons guidance.[1] My experience of the conflict was also "remote" until the summer of 1999, during our family visit to Hvar, when a group of renegade army reservists opened fire on the little stone house my grandfather had built, with us inside it. We crouched behind mattresses pushed up against the walls. Nobody was injured, but we were shaken. After that, my perception of Zaraće changed, emotionally and elementally. Gazing up from the middle of the bay, I wondered whether I could learn to see through the billowing clouds and the refracted blue light to witness the other phenomena of the sky. Sometimes the clouds of *Bura* looked like anvils, forms I had only seen before in cartoons.

In 2017, when I was beginning to conceptualise the shape of this book and Croatia became the first of the former Yugoslav states to progress through the stages of becoming a European Union member, my mother, Diana, was sending me daily pictures of the clouds from Zaraće. At the time, I was reminded of art historian and activist Lucy Lippard's writing on *lures of the local*. Lippard uses the *lure* to describe 'the pull of place that operates on each of us, exposing our politics and our spiritual legacies'.[2] The lure of the local is not only nostalgia for place-based existence or a stronger link to land and community. It is also an argument about the status of place in weathering and responding to regional politics, cultural differences and the global reach of capital. Lippard conveys these paradoxes of the local through her stories of growing up in a small community on the coast of Maine in the 1930s. A parallel story and argument might be told of Zaraće. However, for me, Zaraće's lures are elemental. As I checked my smartphone for new cloud photos, I wondered: what is it that pulls, moves and attracts us to the site-specific and ephemeral expressions of air, atmosphere and weather? How are these feelings informed by history, culture and politics? For Megan Prelinger, the atmosphere holds 'a cultural legacy of missteps and discoveries' that 'press down on us from above'.[3] To counter the airborne incursions at work behind a 'curtain of transparency', she writes, '...our idea of the sky must be reconciled with what we do in it. Our actions there mapped and charted. The sky itself must be woven into the fabric of our societal sense of place'.[4] However changeable, air and atmosphere co-compose the politics of place and planet. The clouds of Zaraće haunted my reflections, congealing one moment, and dissipating the next.

This book responds to the winds of the restless Balkan plateau, the shape-shifting clouds of Zaraće, and the expressiveness of aerial media in many other forms, by developing a notion of *elemental lures*: the 'pull' of aerial, atmospheric and meteorological phenomena. The potential of elemental lures, I suggest, is to reconcile our sensing of air with the myriad social, cultural and political forces suspended in it. It is also to position human sensing of atmosphere in a metaphysical universe of more-than-human sensuality and feeling. Like Lippard, I turn to art, and more specifically the *aerosolar arts*, to apprehend and complicate these lures. In this book, art operates transversally to accounts of atmospheric violence and injustice, proposing alternative modes of *perceiving, moving* and *imagining* (in) the air. Yet the project of *elemental lures* is wider than this. It is about shared rhythms of breathing. It is about tracing the wind as matter moved by the sun, responding to weather in its different figurations, and imagining radically new shapes for the clouds.

Notes

1 I am indebted to my friend and colleague Sophie Dyer for informing me of the notion of 'risk transfer' in conflict. As Eyal Weizman writes in *The Least of All Possible Evils*:

> The conception of risk is central to the calculation of proportionality, especially when attempts to minimize civilian casualties is measured against potential harm to soldiers. The 'trade-off' of risk means that reducing risk

to the attacking military tends to increase the risk to civilians. One of the clearest examples for this 'risk transfer war' was NATO's bombing of Kosovo and Belgrade in 1999. This was mainly due to the decision to conduct high altitude aerial attacks that reduced the danger to NATO air forces, but dramatically increased it for civilians on the ground. The result – no combat fatalities among NATO forces compared with five hundred civilians killed by the bombardment – was understood by many international law scholars as an indication of a breach of the proportionality principle.

(Weizman, 2011: 14)

2 L. Lippard, *The lure of the local: Senses of place in a multicentered society* (New York: New Press, 1997), p. 7.
3 M. Prelinger, 'Charting the sky', in: A. Balkin (ed.), *The atmosphere: A guide* (San Francisco, CA, 2013), p. 1; in suggesting that the atmosphere presses down on us, Prelinger echoes the words of Karl Marx, speaking on the 14th of April 1856 in London: "the atmosphere in which we live weighs upon everyone one with a 20,000-pound force, but do you feel it? No more than European society before 1848 felt the revolutionary atmosphere enveloping and pressing it from all sides" (Marx, 1978: 577). In Prelinger's formulation, however, it is not a 'revolutionary atmosphere' but a history of asymmetric intervention and exploitation that unevenly 'presses' and 'weighs' on the bodies of breathers.
4 Prelinger, 'Charting the sky', p. 1.

References

Lippard, L.R. (1997). *The lure of the local: Senses of place in a multicentered society*. New York: New Press.

Marx, K. (1978 [1856]). Speech at the anniversary of the people's paper. In R.C. Tucker (ed.), *The Marx-Engels reader* (2nd edn, pp. 577–578). London: Norton.

Prelinger, M. (2013). 'Charting the sky'. In A. Balkin (ed.), *The atmosphere: A guide*. San Francisco, CA. Available at: http://tomorrowmorning.net/atmosphere

Weizman, E. (2011). *The least of all possible evils: Humanitarian violence from Arendt to Gaza*. London: Verso Books.

1 Introduction

Elemental lures

I Salt lake

While sitting in a tent at the edge of the Salar de Uyuni in the Bolivian Andes and keeping watch over a rotating telescope recording the movements of astronomical entities hidden in the pale blue sky, I had plenty of time to observe the clouds. They drifted over the horizon of the 10,582 km^2 salt flat, which shimmered in white heat during an unusually dry January. At times, forked tendrils of lightning sparked among and between them. They picked up dust with yellow and red hues. They looked oddly crystalline, like the tessellated lakebed and the salt on the surface of my skin.

I travelled to the Salar de Uyuni in early 2016 with artist Tomás Saraceno and colleagues Tato Chavez, Jan Hattenbach, Maximiliano Laina, Tobias Lange, Bernd Pröschold, Daniel Schulz and Jol Thoms. As we climbed the roads into the Bolivian Andes, we gradually adjusted to the elevation: 3,656 m above sea level. We passed the Lithium extraction pools, where mud is evaporated to produce a light, reactive element prized for its capacity to make currents flow through rechargeable batteries and, paradoxically, to calm flows of excitable neurotransmitters in the human brain.[1] In addition to astronomical filming, we had travelled to the Salar de Uyuni to repeat an experiment that Tomás Saraceno had been performing there since the 1990s.[2] We were preparing to launch *aerosolar sculptures* – pneumatic envelopes that float with the energy of the sun and the movements of air – over the mirror of the salty lake. Due to the lack of rain, however, the lake was no mirror. The radical differences between the sphere of the cosmos, encompassing the clocklike appearance of planets, nebulae and stars, and the sphere of the weather, all perception, movement and imagination, underscored everything we did. Yet in this encounter between astronomy and meteorology, we were only superficially rehearsing responses to changeable environments through the omens of stars, winds, planets and clouds: responses that had far less meaning for us as transient visitors to the Salar de Uyuni than for those making their living with and from it.

Inspired by my visit to the Salar de Uyuni and many similar experiences, in this volume I explore the *elemental lures* of air and atmosphere through practices and experiments of art. Lures are feelings of push, pull and at-traction.[3] They are movements, albeit unruly and unpredictable. Like the crystalline clouds of the Salar de Uyuni or the winds of the Balkan plateau, elemental lures are the expressions of air and atmosphere that attract, unset-tle and reorient us. Yet, to be lured by clouds or winds is not only to perceive aerial phenomena but also to apprehend the cultures and politics of the air, as elaborated in the Preface and in the parable that begins this chapter.[4] In these pages, I illustrate elemental lures by narrating practices and ad-ventures in the *aerosolar arts* in which I have participated over the past six years, at first during an ethnography of Studio Tomás Saraceno in Berlin, and then extending beyond it. Gathered under the names *Museo Aero Solar*, *Becoming Aerosolar* and *Aerocene*, practitioners make, launch and fly aero-solar sculptures as vehicles for questioning the links between aeromobility, advanced capitalism and fossil fuel extraction. Honing techniques of form finding, distributed sensing, weather forecasting, waiting and releasing, the aerosolar arts seek a post-electric[5] imaginary of nomadic movement and elemental intuition.

The lures of art are well documented and theorised[6]; however, their re-lationships to the aerial elements are less so.[7] This volume investigates how art responds to and intervenes in the elements. To think about the elements is to invoke multiple cosmologies, not necessarily beginning, as is often the case in the western academy, with pre-Socratic thinkers like Empedocles. In common usage, the terms 'element' or 'elemental' are multifaceted: they describe different ontological categories of matter, especially air, earth, wa-ter and fire; the environmental milieus and media in which life is immersed; and the scientifically classified substances organised in the Periodic Table. As many have noted, the elements express something tangible and compel-ling about the world while also remaining excessive of human agency and control.[8] They are also useful 'metaphysical descriptors' for thinking about 'phase transitions' of matter and energy under colonialism and empire.[9] For others, to address the 'force of the elemental' is to attend to the world's var-iability and contingency with recourse to the material imagination.[10] This volume is particularly invested in the element of air. The phenomena of the air, from clouds and winds to transmissions and radiation, lure my concep-tual and empirical insights. The aerosolar arts amplify these lures, enlarg-ing spaces for thinking and feeling the air. At the same time, the aerosolar arts modify perceptions, inspire movements and animate the aerial imagi-nary otherwise.

What are the qualities of elemental air? According to Luce Irigaray, "Air does not show itself... it escapes appearing as (a) being".[11] Air, the medium known in opposition to the solid ground or the liquid ocean, moves into our lungs, pores, words and speech, troubling our assumptions of what 'mat-ters'. Air escapes rational description and eludes the visual. Apprehending air as a heuristic for a spectrum of experiences – including those of wind, odour, light, heat, dust, humidity and precipitation – links the personal to the political and the particular to the universal.[12] In other words, to be affected by air, to be a *breather*, is to be in common with other breathing bodies and to register a medium that ties the body to the city, the region and the planet. The aesthetics of air – the way it is sensed in embodied, affective and emotional registers – influences our capacities to confront it as a political problem when pollution plagues our neighbourhoods and monsoons 'condition life and the ability for life to exist as we know it'.[13] In this book, I engage the aesthetics and politics of air by considering, for example, how an artistic experiment at the site of a former military airfield reveals the political and legal 'weather' hovering invisibly in the air. Building on previous work, I also engage the elemental force of air by attending to the relations between the air and the sun.[14]

Thinking about the relations of aesthetics to politics links air to *atmosphere*, a term that has received significant attention in the social sciences in part because it 'connects... the *affective* as a field of potentially sensed palpability with the *meteorological* as the variation in the gaseous medium in which much life on Earth is immersed'.[15] Although some work has presented atmosphere as an abstract metaphor for the transmission, circulation and spatialisation of affect, I foreground the breathable, meteorological and climatic properties of atmospheres as intrinsic to their affective potential.[16] Like Timothy Ingold, Peter Adey and Derek McCormack, I approach atmospheres in a materialist sense by attending to airy-elemental spacetimes unfolding in the midst of, and propagating beyond, bodies, devices and surfaces. In conversation with McCormack who employs an 'entity-oriented ontology' to engage the emergent properties of 'atmospheric things', I explore how atmospheres elicit interest, catalyse movement, stir the imagination, and otherwise lure bodies, entities and matters into novel configurations.[17] In other words, without turning away from a focus on the atmospheric materialisation of entities, I am interested in what ripples between and through atmospheric spacetimes. I am interested in atmosphere as an active, creative and sometimes uncanny force in the composition of worlds. In some cases, I suggest, the lures of air and atmosphere may require a suspension of the idea of the object or entity and an untethering from coordinates of space and discipline.

This volume makes three interrelated gestures. First, I demonstrate the vital role art has played, and will continue to play, in the theorisation of air and atmosphere in the social sciences. Second, I elaborate *elemental lures* to describe the sensual and imaginative propositions of aerial and atmospheric phenomena. If they take shape in winds or clouds, elemental lures do not collapse into romantic appreciation of 'Nature'; rather, these winds and clouds also heighten our attentions to social, cultural and political conditions. Finally, I narrate experiments in the aerosolar arts to trace novel modes of *perceiving*, *moving* and *imagining* (in) the air. The remainder of this chapter contextualises this volume's contributions. In the next section, I introduce the aerosolar arts in greater detail and outline the methods that informed my research and practice. Then, I attend to the role of art in social scientific approaches to air and atmosphere. In doing so, I position art at the centre of debates on material ontologies and imaginaries. After a discussion of the role of art and the aerial elements in feminist approaches to politics and difference, this introduction concludes with an outline of the chapters to come.

II Aerosolar arts

This volume draws on ten years of research into the aesthetics and politics of air through the lures of art, beginning in my undergraduate thesis on Olafur Eliasson's spatial experiments in 2010, and extending through my current work as an aerosolar practitioner and co-organiser (with Sophie Dyer) of the feminist project Open Weather. The empirical work featured in this book begins in a creative ethnography of Studio Tomás Saraceno in Berlin that formed a central part of my doctoral research. Studio Saraceno was established by Tomás Saraceno in Frankfurt in 2005. Over time, it grew into a multidisciplinary practice focused on the structures and poetics of clouds, the webbed worlds of spiders, and more generally, the design of ecologies for living and moving on, and above, the Earth. Saraceno and his team were the first to scan and model a black widow spider's web in three dimensions. In an ongoing body of work called *Cloud Cities*, studio members experiment with pneumatic forms and Weire–Phelan geometries toward a vision of nomadic aerial habitats. Parallel projects have explored vibration in interspecies communication via collaborations with bio-tremologists, and the politics of imperceptible forces and materials, from black carbon to cosmic dust. I first encountered Saraceno's work in 2012 on the roof of the New York Metropolitan Museum, where a steel and mirror *Cloud City* sculpture reflected the sky and the colours of trees in Central Park.

In the same year, the studio moved from Frankfurt to Berlin and spent a brief interim in a space near the Hauptbahnhof before moving into the former administrative building of Actien-Gesellschaft für Anilin-Fabrication (a dye company that manufactured photographic film) in Berlin-Rummelsburg. By chance, I met Emek Ulusay, a core member of Studio Saraceno, at a symposium on Saraceno's installation *On Space Time Foam* (2012–2013) at the Hangar Bicocca in Milan and was invited to the studio as its first doctoral researcher in early 2014. There, I was introduced to a lively 'ecology of practices' involving the production of artworks, exhibitions, publications and pedagogies.[18] I witnessed members of the studio carry out experiments with South American spider species, Riemannian geometries and stratospheric balloons. My planned three-month field visit extended further and further. For the following three years, I collaborated with Tomás and many practitioners at Studio Saraceno (numbering around twenty when I arrived, and fifty when I left, while spider populations increased to thousands). I became 'mixed up' with events in Germany, France, Austria, USA, UK, Bolivia and Turkey. This included participating in exhibition conception and production, collaborative writing and presentation, modelling and fabricating sculptures, and teaching (including the design and delivery of a two-year curriculum for architecture-related art). Despite these engagements, I can neither claim that I have an objective understanding of the studio environment, nor assert that my ethnography was comprehensive. It was never my explicit goal, for example, to develop a network-based analysis of studio society, to document with a microscope the studio's economic flows, or to explore hierarchies of studio management and its relationship to other institutions.

Instead, I investigated the studio's 'expanded field' of artistic practices through a range of methods including participant interviews and observation, focus groups, creative writing, image-making and diagramming, through which I recorded many stories of creative projects realised and unrealised, concepts tested and abandoned, and the personal histories caught up in them.[19] In the process, I came to understand my methodology as 'tracking' or 'feeling-into' a specific current of work at Studio Saraceno: the current of the aerosolar arts.[20] In my doctoral research, this approach enabled an attention to flows of knowledge, percolating ideas, travelling concepts and blurred borders of theory and practice. It had an additional consequence that after I completed my dissertation and left Berlin, I continued to think, move and practice with the aerosolar arts. At the same time, these aerosolar practices and their networks extended further outward from Studio Saraceno into many other cultural and institutional spaces. While I remain acutely aware of its beginnings, the majority of work presented in this volume can no longer be called an ethnography; rather, it is an archive of participation in the hybrid practices of three aerosolar initiatives that were inspired by Tomás Saraceno and members of his studio: *Museo Aero Solar, Becoming Aerosolar* and *Aerocene*. Through narrating each of these projects, I investigate the role of art in shifting perceptions, catalysing movements and enlarging alternative imaginaries of air and atmosphere.

There are many reasons why the aerosolar arts are fitting companions for this inquiry. In the aerosolar arts, the air and the sun form primary materials. Using techniques of design, crafting, modelling, recycling, sewing and hacking, aerosolar practitioners produce pneumatic, balloon-like forms that can fly using only the sun and the air. Although *Museo Aero Solar, Becoming Aerosolar* and *Aerocene* are the particular instantiations of the aerosolar arts featured in this volume, they belong to a wider historical field of experimentation with aerial and solar aesthetics and politics. This wider field would include solar balloon designs by twentieth-century inventors like Tracy Barnes, Dominic Michaelis and Frederick Eshoo; the spectacular *baloeiros* of Brazil; the pneumatic and political demonstrations of Ant Farm; the first crossing of the English Channel in a solar aerostat by Julian Nott in 1981; the airfields of Graham Stevens; the Montgolfière Infra-Red (MIR) balloon of the Centre Nationale des Études Spatiales (CNES); advanced solar aerostats made by prize-winning secondary school students Axel Talon and Etienne Lalique; and the myriad forms of inflatable and solar-sensitive protest art by Tools for Action, among others. Working with a growing team at his studio, Tomás Saraceno has become a leading advocate and inventor of solar-aerostatic practices.

As artistic interventions, *Museo Aero Solar*, *Becoming Aerosolar* and *Aerocene* enter this existing aero- and heliosphere of design, architecture and experiment. *Museo Aero Solar*, the oldest of these three projects, was established in 2007 by Tomás Saraceno and Alberto Pesavento. It evolved as a community as well as a collectively fabricated aerosolar sculpture composed of reused plastic shopping bags cut into rectangular shapes and connected into a membrane to be inflated with air and to absorb solar energy. True to its name, *Museo Aero Solar* is a flying museum and an archive of plastic matter, consumption patterns and urban ecologies. As I learned from interacting with Tomás Saraceno, Alberto Pesavento, Jatun Risba and Till Hergenhahn, the *Museo Aero Solar* community fluctuated in membership, sites of activation and core principles. My interaction with this project began in 2014 when Tomás Saraceno invited me to participate in *Museo Aero Solar* activities at the Les Abattoirs museum in Toulouse. Over time, and together with friends and colleagues, I began initiating *Museo Aero Solar* workshops in the UK. Dozens of other workshops have occurred elsewhere. While these events speak to each other across time and space, and are shared among aerosolar practitioner networks, the project's politics has

Figure 1.1 Museo Aero Solar, 2007–ongoing; at Prato, Italy, in 2009 with Alberto Pesavento, Tomás Saraceno, Janis Elko, Till Hergenhahn, Giovanni Giaretta, Marco Alessandro, Manuel Scano, Michela Sacchetto and Matteo Mascheroni.

Note: Initiated by artist Tomás Saraceno in conversation with Alberto Pesavento in 2007, *Museo Aero Solar* unfolds in the space formed between human and nonhuman participants in the simple acts of cooperation and reusing plastic bags, to collectively produce an aerosolar sculpture. Fostered in more than twenty-one countries to date, *Museo Aero Solar* embodies a vision of pollution-free futures through the growth of self-assembling, geographically dispersed participatory communities; in this way, the practice can be seen as marking the beginning of the genealogy of *Aerocene*. Courtesy *Museo Aero Solar* and *Aerocene* Foundation.

Source: Photography by Janis Elko. Licensed under CC BY-SA 4.0.

continually altered and evolved. As I show in Chapter 2, *Museo Aero Solar* is taking on new life in Buenos Aires, where, since late 2017, several new sculptures have been constructed and, more importantly, connected to a series of political struggles in the city and its suburbs.

As the composition and character of *Museo Aero Solar* shifted, another aerosolar project gained momentum. *Becoming Aerosolar* was a concept and series of aerosolar practices launched in the summer of 2014 that not only borrowed tools from *Museo Aero Solar* but also extended beyond it into other interdisciplinary inquiries. A text published in the volume *Art in the Anthropocene* explains:

> To *become aerosolar* is to imagine a metabolic and thermodynamic transformation of human societies' relation with both the Earth and the Sun. It is an invitation to think of new ways to move and sense the circulation of energy. And, it is a scalable process to re-pattern atmospheric dwelling and politics through an open-source ecology of practices, models, data—and a sensitivity to the more-than-human world.[21]

The project drew inspiration from the hypotheses of Nikolai Kardashev and Bronislaw Szerszynski in its proposal of a future 'solar-cene' in which societies would be liberated from the Earth's surface by harnessing the circulation of energy in the air. Taking shape in a range of projects, sculptures and collaborative writings, *Becoming Aerosolar* was further developed at the Institut für Architekturbezogene Kunst (IAK) at the Technical University of Braunschweig, where Tomás Saraceno served as the director for two years and Ivana Franke, Natalija Miodragovic, Alan Prohm, Jol Thoms and I worked as lecturers. In my view, *Becoming Aerosolar* cannot be dissociated from the work of the 300 undergraduate and postgraduate students who attended courses at IAK. The *Becoming Aerosolar* curriculum featured the theory of pneumatic architecture, aerosolar sculpture workshops, aero-geography, tensegrity modelling, and the invention of forms of care for IAK and its local environment. As I will elaborate in Chapter 3, students contributed ideas for aerosolar societies, honed prototypes for 'cut-down mechanisms', drafted the first *Manual for Becoming Aerosolar*, produced work for international exhibitions and significantly evolved the dialogue on aerosolar initiatives together with more experienced practitioners.

The *Aerocene* project is the most recent manifestation of the aerosolar arts explored in this volume, and the one I am most entangled with as both a researcher and practitioner. Initiated by Tomás Saraceno in 2015 with the launch of the first fully certified, human-carrying solar balloon – the *D-OAEC Aerocene* – at the White Sands National Monument in New Mexico, *Aerocene* has grown into an international network of practitioners from Berlin to Buenos Aires to London. Among many other achievements,

Figure 1.2 Becoming Aerosolar, 2015; aerosolar sculpture launch in Wolfenbuttel, Germany with Tomás Saraceno, Jol Thoms, Alexander Bouchner, Henry Kirchberger, Lok Junlin Luo, Jehona Nuhija, Tomi Šoletic, Karla Sršen, Bruna Stipanicic and Ananda Wiegandt.

Source: Photography by Sasha Engelmann, 2015.

in recent years *Aerocene* has collaborated with Lodovica Illari, Glen Flierl and Bill McKenna of the Earth, Atmospheric and Planetary Sciences department at the Massachusetts Institute of Technology to develop the *Aerocene Float Predictor*, a tool to forecast the round-the-world movements of floating aerosolar sculptures using National Oceanic and Atmospheric Administration (NOAA) wind data.[22] The *Aerocene* project has been exhibited internationally and was featured in Tomás Saraceno's carte blanche exhibition *On Air* at the Palais de Tokyo in 2018. More recently, *Fly with Aerocene Pacha* was an action led by Tomás Saraceno as part of the global art initiative Connect-BTS, supported by the famous Korean-pop band BTS. In Jujuy, Argentina, the *Fly with Aerocene Pacha* team achieved the first ever autonomous air travel with the *D-OAEC* Aerocene sculpture piloted by Leticia Marques. In solidarity with indigenous communities from Tres Pozos, Pozo Colorado, San Miguel del Colorado and Inti Killa, the sculpture carried a collectively written message that read 'El agua y la vida valen más que el litio / Water and Life are worth more than Lithium'. The action has since seeded a growing (albeit unusual) coalition between local environmental activists, BTS fans and Aerocene practitioners in the region.

Although *Aerocene* is rightly called a community, and it depends on relationships among dozens of people with different forms of expertise, we are still developing a working model of community governance. This is an active area of discussion among members. It also means, as is often the case in hybrid communities, that many collaborators, research assistants and interns have lacked platforms to publicise their contributions. In claiming a modest role in the project, I must emphasise that my work has always evolved in dialogue with many other practitioners. For example, since *Aerocene*'s early days, my participation has involved developing educational tools and pedagogies with *Aerocene* sculptures for academic and public-facing venues. Together with many fantastic colleagues, I also co-curated *Aerocene* Symposia at the Haus der Kulturen der Welt in Berlin, the Royal College of Art in London and the Palais de Tokyo in Paris. At the time of writing, I am collaborating with Argentine colleagues Débora Swistun and Joaquin Ezcurra, as well as a team of engineers and earth scientists at Royal Holloway University, to develop air quality monitoring capacities for Aerocene sculptures, and to explore equity based models of environmental sensing with communities on the outskirts of Buenos Aires.

Although I make distinctions between *Museo Aero Solar, Becoming Aerosolar* and *Aerocene*, these currents of the aerosolar arts are mutually permeable. For example, as will be elaborated further on, the *Becoming Aerosolar* curriculum at IAK included the fabrication of *Museo Aero Solar* sculptures. The *Museo Aero Solar* sculpture constructed and launched in 2014 in Lima, Peru on the occasion of the Development and Climate Days conference has sometimes been called *Becoming Aerosolar*. Since 2014, *Museo Aero Solar* has frequently been exhibited with *Aerocene*. Sometimes the works of *Museo Aero Solar* and *Becoming Aerosolar* are presented in exhibitions as historical instantiations of *Aerocene*. This book is not a distanced account of these three strands of the aerosolar arts. Rather, from my position within these collective spaces, I trace how they intervene in air and atmosphere and *lure* bodies into different forms of elementally engaged practice and politics.

Figure 1.3 Tomás Saraceno for *Aerocene*, 2016; the photographs were taken at Salar de Uyuni, Bolivia with the support of the Barbican Art Gallery, London.

Source: Courtesy the artist; Tanya Bonakdar Gallery, New York; Andersen's, Copenhagen; Pinksummer contemporary art, Genoa; Esther Schipper, Berlin. © Photography by Studio Tomás Saraceno, 2016 licenced under CC BY-SA 4.0.

Art has already made significant contributions to the theorisation of air and atmosphere in the social sciences. To better grasp how and where the aerosolar arts contribute to this body of work, the following section outlines several ways in which art has informed geographies of air, atmosphere and the elements. Importantly, the role of art is not reducible to the illustration of academic concepts or intellectual arguments. Artworks enlarge forms of aesthetic and political perception of the air; they inspire movements across space, text and discipline, and they activate the elemental imagination through representation, figuration and allegory.

III Aërography

Aesthetic sensibilities have been part and parcel of geographical attentions to the elements. We can begin to trace these attentions by highlighting 'sensual intelligibility' in nineteenth- and early-twentieth-century geographers' engagements with elemental matters.[23] To make a brief tangent away from air: the geomorphologist G.K. Gilbert employed 'geologic cross-sections, field sketches and photography, as well as... chromolithographics [and] woodcuts' to render landscape 'sensually intelligible' in what some have called 'topographic art'.[24] These investments persist in contemporary collaborations between geomorphologists and artists that have been recognised as intrinsic to the health of the discipline.[25] Returning to elemental air, in 1916 Alexander McAdie called for the establishment of *aërography*: the study of the description of the air with emphasis on its relationship to life.[26] This 'air-writing' would complement *aërology* or the study of 'the whole domain of atmospherics' at the planetary scale.[27] McAdie emphasised the importance of descriptive charts of atmospheric layers to aid the nascent 'art of aerial navigation'.[28] In an echo of Gilbert's topographic arts, McAdie and his contemporaries attempted to overcome the geographical 'handicap' to flat surfaces by rendering the atmosphere as *sensually intelligible* as the Earth. Although McAdie's contributions most directly informed physical geography and climatology, his proposals for aerial description anticipated more recent phenomenologically informed attempts to 'rematerialise' the discipline of geography by unearthing it.

We can further trace aesthetic sensibilities of the elements in the notion of *lures for feeling*, which appears in more recent work on geography's material imaginary. Ben Anderson and John Wylie call upon 'the properties of any element (ie earth, wind, fire, air) and/or any state (ie solid, liquid, gaseous)' to untether geography's material imaginary from ground and solidity.[29] They elaborate:

> Our aim... is to experiment with different material imaginations that are efficacious not because they offer a conception of the world, one that functions through an effect of adhesion, but because they aim to elicit interest in that 'they act as a lure for feeling, for feeling that "something matters."'[30]

Anderson and Wylie invoke elemental imaginaries not because they explain the world, but because they 'elicit interest', citing Isabelle Stengers' use of the Whiteheadian expression *lures for feeling*. The phrase *lures for feeling* originates in Alfred North Whitehead's claim that theories and imaginaries, or 'propositions', are not relevant or interesting because they present the truth. Rather, they elicit different degrees of feeling: they act as lures.[31] According to Whitehead, who was a mathematician turned philosopher, lures for feeling may be felt by audiences in the theatre as well as by 'imaginative historians' speculating on alternative outcomes to past events.[32] For Whitehead, imaginative flight and intellectual adventure are equal ways of 'feeling the proposition'.[33] Thus, on one level, Anderson and Wylie's project is to test imaginaries and theories of the elements that may become 'interesting' lures for geographical thought. In tracing a brief history of aero-geographical aesthetics and artistry, it is significant that lures, rather than optical metaphors or rational tools, play a primary role in transforming geography's material imaginary.

However, more analytically, propositions or lures are independent from human intellect; they are also interesting and alluring to the non-human, atomic and ephemeral units of reality that Whitehead calls 'actual entities'.[34] As Melanie Sehgal expands, 'propositions... operate on the preconscious, metaphysical level of feeling'.[35] In other words, propositions produce ripples in the material world. Sehgal elaborates, 'An explicit thought or phrase is the outcome of the tremendous excitement that propositions induce, if they successfully "lure a feeling"'.[36] In this formulation, the lure of particular material concepts and imaginaries is not entirely, or even primarily, the result of human intellectual assertion; rather, interest in these concepts and imaginaries *presupposes propositions*, or as McCormack translates, 'the alluring call of the world as it makes the actuality of its becoming felt differentially'.[37] Thus, Anderson and Wylie describe a variety of 'alluring' elemental propositions, from 'the glitter of liberated ore' to 'the writhing jet stream', to suggest how geographers' material imaginaries might be lured by a capacious series of propositions at play in the world.[38] Moreover, in doing so, these authors marshal many properties associated with air, including 'turbulence', 'precipitation' and 'airiness'.[39] Air makes compelling propositions: palpable lures for feeling. Although Anderson and Wylie do not cite many artworks, their recognition of elemental lures in the geographical imagination and their prominent use of aerial metaphors set the scene for later contributions.

As many have noted since the publication of Anderson and Wylie's article, perhaps more so than any other element, geography is haunted by air. It is true that similar statements could be made about water, and indeed many geographers have turned to the wet, watery and liquid.[40] However, air's quasi-invisible properties and its common associations as solidity's 'opposite' or 'lack' make it a particularly interesting place to start for querying and re-imagining geography's material foundations. Air poses a challenge to geography in that, to start with air, as aërographers like Mark Jackson and Maria Fannin advocate, is to let go of fixed spatial or scalar lenses. It is to *suspend* disciplinary frameworks and positions from the very start. To follow air inside and outside the body, through the pores of infrastructure and through the interstices of the map requires practical and conceptual creativity. Jackson and Fannin ask: 'where would our material geographies take us then?'[41] In other words: *where would air lure* us? How would it move us? Which 'new imaginative structures for thought or politics' would emerge along the way?[42]

In addressing this question – *where would air lure us?* – the articles in a special issue on aërography feature discussions of the 'sky art' of James Turrell,[43] the poetry of Paul Celan,[44] the installations of HeHe (Helen Evans and Heiko Hansen),[45] the 'haunted sound' of John Cage,[46] the theatrical masques of Ben Jonson and Inigo Jones,[47] the turbulent worlds of Louise Erdrich's *The Painted Drum*[48] and the air architecture of Yves Klein and Robert Barry.[49] When investigating methods and practices for beginning with air, art and literature are seemingly inescapable. However, the role of art in this special issue is not only to expand the repertoire of aerial description through different media. Rather, art contributes to the querying of ontological frameworks that privilege solidity, the object and the ground as starting points for research and as platforms for the subject. For Yuriko Saito, art also provokes the constructed nature of disciplinary knowledge systems and their inheritance of western legacies of thought that foreground Being – 'an independent, discrete and permanent substance' – over notions of contingency and emptiness from eastern traditions.[50] Across the eight articles of the special issue on aërography, works of art and literature respond to and engage with the material propositions, or lures, of the air.[51] In doing so, these artworks also function as imaginative, intellectual and bodily lures in their own right. They lure readers and audiences to the sensuous materiality of air, the suspension of history and memory, the entanglement of bodies and atmospheres, and an awareness of the tacit philosophical premises of geographical thought.

More precisely, however, how do works of art function as *lures* in aerographic writings? To better grasp the lures of art, we need to investigate what is it that art renders *perceptible*. Let us consider an example. Peter Adey's scholarship engages with historical fiction to apprehend the elemental qualities of air. In order 'to attune with far more precision air with the elemental', Adey studies the literary geography of Johann Goethe's *Elective Affinities*: 'a chemical landscape told through airs of appearance, attraction and repulsion, new pairs and bondings'.[52] Scenes from Goethe's novel convey an older sense of elemental air with connotations of bodily humours, morality and the hormones of the imagination.[53] However, a critical geography of elemental air, Adey suggests, would 'listen' to stories of class and gender as well as power and difference as they are implicated in air's felt qualities. Adey demonstrates this through exploring 'how elemental air gains shape in writing'.[54] More specifically, he reads Goethe alongside Elizabeth Gaskell's *North and South* and Charles Dickens' *Little Dorrit*, arguing:

> The accounting of air in writing like this enables us to distil the inequality of labour relations, working and living conditions through the

elemental. Air penetrated language and gave shape to Dickens, Gaskell and others making sense of an emerging public health discourse and the bigger machinic, economic and spiritual forces at play in the fates of their characters.[55]

In these novels, then, air becomes much more than its aesthetic expression; it is simultaneously a vector of relations of power in material, moral and spiritual guises. The works of Goethe, Gaskell and Dickens convey air's phenomenological qualities while amplifying its politics. Adey's reading of these novels demonstrates that art lures our perceptions of air as simultaneously aesthetic and political. The role of art in crafting *lures of perception* of air and atmosphere will be further elaborated in Chapter 3.

Art has also been fundamental in nuancing perceptions of the weather. Hence, we can turn to cultural climatology, where geographers like John Thornes, writing on the paintings of Monet and Constable, 'make visible the meteorology at work in popular representations of the sky'.[56] Thornes and others interpret the hues, textures and techniques employed by visual artists to 'deconstruct proxy data' of weather and the ranges of pollution present at the time an image was created.[57] In turn, these paintings become forensic documents of historical atmospheres. Equally interesting is Thornes' treatment of Olafur Eliasson's *The Weather Project*, presented as a continuation of Monet's *London Fog* series and an invitation to involve citizens in the production of weather forecasts.[58] In other words, *The Weather Project* indicates the potential of involving audiences in the making, circulating and interpreting of airy-elemental experiences in a kind of crowd-sourced cultural climatology.

In addition to considering what art renders perceptible in our engagements with air, we might also ask how art *moves* citizens, scholars and readers. A lure for feeling is a 'vector' since it 'feel[s] what is *there* and transform[s] it into what is *here*'.[59] As I suggested before, a lure is a kind of pull or push. If art has the capacity to lure, where does art take us that is different from where we started? Thinking with Beatriz da Costa's project *PigeonBlog*, a collaborative initiative between homing pigeons, artists, engineers and pigeon fanciers engaged in a grassroots data-gathering effort about air quality in Southern California, Georgina Born and Andrew Barry discuss *epideixis*, or 'the transformative power of speech and art, a *power to move*' beyond the proof or demonstration.[60] The movement described by Born and Barry is of air, human and nonhuman (pigeon) bodies as well as the 'relations between new knowledge, things, locations and persons that did not exist before'.[61] As sensor-equipped pigeons travel through the folds of air in Los Angeles, mapping the flux of air pollutants, this public aerial experiment proposes new movements through space and discipline.

Several other scholars have traced air's movements through forms of writing that invite air *in* to the surface of the page. For example, Timothy Choy's writing on 'air's substantiations' manifests in experimental ethnographic insights about air in Hong Kong, including excerpts from short stories of floating by Xi Xi, and the 'airy poetics' of Shigehisa Kuriyama.[62] On Kuriyama, Choy writes, 'wind and air whistle through his writing as much as they do through the texts he analyses'.[63] Choy finds in the works of writers and poets a sense of air's 'whirlings, its blowing through scales and borders, its condensations, its physical engagements, its freight of colonial, economic and bodily worries'.[64] As I have demonstrated elsewhere via Choy's 'airy poetics' and Dryden Goodwin's *Breathe*, works of art move and reconstitute forms of knowledge in the air, while also moving air into text, surface and body, breathing new associations along the way.[65] These are *lures of movement* that will return to our focus in Chapter 4.

Art has a third important quality as a lure for social scientific scholarship, namely its capacity to access the imagination through figuration, metaphor and allegory. Indeed, Whitehead describes the lure of imaginative propositions as a 'penumbra' of alternatives, whether in relation to events, ideas or decisions.[66] In this vein, Jeffrey Jerome Cohen and Lowell Duckert ask: 'Is there potential in the impossible, in the purely imaginary, in the abandoned and the unreal (ether, phlogiston, the sea above the clouds)?'[67] One answer to this query lies in Steve Mentz's essay on the imaginary combustive element of *phlogiston*.[68] Mentz explores Shakespeare, Edmund Spenser and the writings of early modern alchemists, to trace 'moral knots' between separation and combination, kingdom and house, order and chaos, proving that a fictional element yokes the imagination to particular material properties.[69] In *Aerial Aftermaths*, Karen Caplan provides another possible answer to Cohen and Duckert's question. Caplan investigates Man Ray and Marcel Duchamp's *Élevage de poussiere* (*Dust Breeding*), a photograph of one year's worth of dust on Duchamp's sculpture *The Bride Stripped Bare by Her Bachelors, Even*.[70] Transcending micro and macro, the photograph resembles a strange, barren landscape seen from above. As an experiment in the visual culture of modernist abstraction via the molecules of dust, Ray and Duchamp's image becomes an allegory for the western fantasy of the 'desert' that is 'highly aestheticized, powerfully negating'.[71] This allegorical relationship of geometrical abstraction, aerial conflict and the desert is further complicated in Caplan's writing through an attention to Sofie Ristelhueber's eerie photographs of Kuwait after the Gulf War. Ristelhueber's photographs are 'terrain[s] that have been forcefully and deliberately made' signifying an 'active format' connecting artwork and world.[72] In tracing 'what is folded into images as well as what extends out from them', Caplan enrols the lures of art to explore the wakes and worlds of aerial power.[73]

In other works, art enables a critique of prominent atmospheric imaginaries that obscure relations of power and domination. For example, Louise Amoore engages with Trevor Paglen's photographs to demonstrate that the figuration of 'the cloud' as metonym for cloud computing manipulates the aerial imaginary to blur political asymmetries.[74] Paglen's work makes visible the use of allegory and metaphor to dematerialise cloud computing services and their effects on bodies, borders, landscapes and climates. The ghostly undersea cables and NSA data centres populating Paglen's 'visual vocabulary' counter these dematerialising allegories, while suggesting another powerful imaginative matrix of the deep sea, air, earth, server and algorithm.[75] Yet this matrix does not stop with cables and servers; it also includes formations of finance, state-led spying, corporate communication, internet probes and critical geopolitics. From fictional elemental substances to the clouds of the surveillance state, then, works of art function as *lures of imagination*; through the resources of metaphor, figuration and allegory, these lures invite us to question imaginative tropes of air and atmosphere and to evaluate the relations they authorise.

In previous work, Derek McCormack and I wrote about the aerosolar arts as 'more than... a lure for thought'.[76] For us, the aerosolar sculptures emerging from the practices of *Museo Aero Solar and Aerocene* are 'attractors' that are 'moved, pulled, and pushed by the elemental force of the sun'.[77] To elaborate, aerosolar sculptures are not only intellectually exciting devices for thought; they are also highly responsive to the propositions of solar energy, understood as the animation of the atmosphere: 'the wind we feel is always a solar-powered wind'.[78] At the same time, aerosolar sculptures are dependent on collective practices that constantly adapt to elemental circumstances and intervene in heavily regulated airspaces. Thus, aerosolar sculptures perform the 'force of the elemental' while enlarging spaces for sensing the politics of the air. In their affective envelopments, transboundary movements and imaginative figurations, aerosolar sculptures amplify lures of air and atmosphere. Through a focus on the aerosolar arts, and employing a theory of lures, in this book I further develop art's contributions for scholarship on air, atmosphere and the elemental.

As briefly elaborated in this section, works of art including fiction, poetry, photography, film, sculpture, performance and theatre have informed attentions to air and atmosphere in the social sciences, and more specifically, in geography. Yet as aërographic scholarship shows, the role of art cannot be collapsed into modes of description, illustration or demonstration. Rather, a flourishing body of writing on air and atmosphere reveals that works of art actively pull, push and otherwise move disciplinary ontologies. This occurs via expanded forms of perception, movements through space, text and discipline, and the imaginative forces of figuration, allegory and metaphor. These *lures of perception, movement* and *imagination* are not isolated, as the examples in this section prove. Nor are the lures of art confined to these three modes. However, they offer a means of departure for grasping the role of art in discourses on air and atmosphere. They also establish a framework through which to trace art's lures in case studies, experiments and practices. The testing of this framework occurs in the empirical chapters of this volume.

I have so far engaged in debates on air and atmosphere by foregrounding the propositions of art. However, scholarly and creative works on atmosphere in the social sciences join many other elementally informed projects in other disciplines. For example, Christina Sharpe thinks with Dionne Brand,

M. NourbeSe Philip and Toni Morrison to convey *the wake* and *the weather* of slavery and antiblackness.[79] Macarena Gomez-Barris turns to the photographs and films of Francisco Huichaqueo and Carolina Caycedo to discuss a watery, *submerged perspective*, and to illustrate indigenous practices of world making in the interstices of extractive capitalism.[80] Astrida Neimanis writes with Rebecca Belmore's sculptures and *bodies of water* to elaborate on a posthuman feminist phenomenology.[81] Elizabeth M. DeLoughrey reads allegories of earth, radiation and ocean with indigenous artists of the Pacific, such as Kamau Braithwaite and Kathy Jetñil-Kijiner.[82] These authors engage with decolonial, queer and indigenous artists who explicitly work in and with elemental media. What I take from this scholarship is a powerful affirmation of the role of art in the naming, critique and refusal of forces of racism, sexism and coloniality as they propagate in and through air, water and soil. These authors also signal the value of attending to elemental materials and imaginaries in the analysis of the politics of location. The following section further traces the relationship between the lures of art, elemental air and feminist approaches to location and difference, concerns that echo through the core chapters of this volume.

IV Toward an elemental politics of location

In this book, I reach beyond the social sciences into feminist thought, cultural studies and the humanities, to further trace the important links between art, elemental air, and questions of politics and difference. Like Peter Adey, I find that more critical work on these relationships is needed. In particular, feminist ideas on the politics of location and the ethics of elemental exposure, especially in relation to air and atmosphere, are furthered through artistic practices and resources. Before turning to this work, however, we must recognise the colonial legacies of dominant elemental constructs. This goes beyond the fact that in many dedicated studies, the classical conception of the elements, usually derived from Empedocles and other pre-Socratics, is given precedence as the origin for elemental thinking.[83] As Yuriko Furuhata argues, there is a need to 'expand the referential framework of the 'elements' beyond ancient Greece in order to productively complicate the geopolitics of elemental philosophy'.[84] In addition, many have demonstrated the complicity between specific western elemental ideologies and colonial, racist or sexist projects. To name only a few, scholars have probed constructions of the weather as it informed chattel slavery,[85] theories of the breathing body in racist eugenics[86] and the importance of Beaufort-scale wind measurements in the trans-Atlantic slave trade.[87] The conditions of climate change have been described by Denise Ferreira da Silva and Sasha Langford as the elemental 'phase transition' of colonial expropriation and racial capitalism.[88]

As a white, settler woman, engaging with the elements from a position in the western academy, I am complicit with the forms of institutional injustice outlined above. As an Anglophone geographer, I draw from North American and British discourses that inherit European humanist perspectives. However, in a modest way, and in conversation with interlocutors from feminist and activist traditions, I seek to reframe and destabilise normative accounts of air and atmosphere. In this volume, instead of treating air as an 'outside' or 'object', I probe the different ways bodies are implicated in, and affected by, air and atmosphere. This includes thinking through the sharing of affective-meteorological atmospheres while highlighting the limits of what is held in common. It includes making space for ideas that have not been published in scholarly venues. Through narrating from personal experience and with members of aerosolar communities, I show that the politics of air is inherently malleable, never foreclosed by dominant forces. I also seek to disfigure and complicate images and narratives of aerial exploits that are too quickly interpreted as acts of heroism.

Attending to air and atmosphere in these ways, I learn from a vibrant body of work linking air and water to the decades-old feminist method of the 'politics of location'. Introduced by black feminists like those of the Combahee River Collective and developed by Adrienne Rich, Audre Lorde, Gloria Anzaldúa and Rosi Braidotti among others, the politics of location begins with 'the geography closest in – the body'.[89] From the 'scars, disfigurements, discolorations' to 'the teeth of a middle-class person', Rich's body speaks of intersecting forms of power, privilege and struggle.[90] As both a method and an analytic that begins with the body, the politics of location addresses 'historical, geographical, cultural, psychic and imaginative boundaries which provide the ground for political definition'[91] and involves mapping 'the networks of power in which [our bodies] are situated'[92] in order to apprehend the matrix of domination.[93] Recently, air and water figure prominently in these cartographies. For example, through the work of Colombian artist Carolina Caycedo, Macarena Gomez-Barris traces the relations between social repression, the Magdalena 'river body' and logics of hydropower in Colombia.[94] Expanding beyond the watershed, Astrida Neimanis writes that racism, sexism and colonialism are 'carried by currents in a weather-and-water world of planetary circulation, where we cannot calculate a politics of location according to stable cartographies'.[95] For Neimanis and Denise Ferreira da Silva, among others, we must learn how to trace our bodily politics in the elemental movements of wind, water and weather, in uneven flows of toxins, in the propagations of infrastructure and logistics, and in the thermodynamic thresholds of our planet.

Many scholars and artists are already expanding this project through discursive and creative practices. Christina Sharpe writes of slavery as 'a singularity – a weather event or phenomenon likely to occur around a particular time, or date, or set of circumstances'.[96] In other words, slavery is not singular, as in passed or historical; rather, Black Life is continually 'held' in the singularity, a 'singularity of antiblackness'.[97] If the weather is the condition of time and place, however, 'antiblackness is pervasive as climate'.[98] To elaborate on this point, Sharpe constructs a capacious definition of the climate in quotidian events of police brutality, in migrant vessels leaving Libya and in the aftermath of the Haitian earthquake of 2010. Importantly, antiblack climate for Sharpe is not an atmospheric metaphor but a material condition. For Sharpe, the poems, prose and songs of Kamau Braithwaite, Dionne Brand and M. NourbeSe Philip become barometers and navigational tools for living *in the wake* of slavery and in the weather of racism. These artists' gestures provide ways of observing, mediating and responding to 'insistent Black exclusion' and 'un/survival' in the singularity.[99]

Thinking with Sharpe's figuration of the 'total climate' of antiblackness, Astrida Neimanis and Jennifer Mae Hamilton bring location and difference to bear on the weather through the notion of *weathering*. For these authors, weathering is the process of being affected by physical as well as socio-political weather systems. Asserting that 'weathering is not a metaphor', they elaborate:

> In the face of the greatest climatic transformation that human bodies have ever known, weathering means learning to live with the changing conditions of rainfall, drought, heat, thaw and storm as never separable from the 'total climate' of [the] social, political and cultural existence of bodies.[100]

Neimanis and Hamilton argue that forces of colonialism, racism or sexism are not 'like' or 'analogous to' the weather; instead, these forces *are weather*. Weathering makes us responsive to our uneven exposures to these 'more-than-meteorological' weather systems. Hamilton and Neimanis' *Field Guide for Weathering* employs speculative questions, embodied experiments and a 'cosmic weathering meditation' to expand weathering repertoires for artistic and academic communities.[101] For these practitioners, weathering requires creative acts of corporeal awareness, as well as the diagnosis of power suspended in the elements.

Probing the relationship between the politics of location and the 'weather-and-water world' is an underlying motif in this volume. Through the lures of the aerosolar arts, I engage some of the approaches expanded by the previously cited authors. Aerosolar artworks amplify an elemental politics of location by foregrounding personal histories, corporeal capacities, and by acting as wayfinding instruments in the heavy weather of the present. At the same time, as the aforementioned works reveal, the role of art is not necessarily to further narratives of damage. To engage with art is to test forms of expression that do not easily collapse into categorisations of harm, illness and vulnerability. In this way, art can offer vital resources for addressing the violence suspended in the air without reinforcing the old categories and power axes that produced these conditions in the first place. The crafting of alternate narratives of aerial life is one of the potential contributions of the art featured in this volume. The remainder of this chapter provides an outline of chapters to come.

V Chapter outline

Through a series of empirical investigations and narrations of art, this book advances the *lure* for elemental aesthetics and politics and, in the process, contributes to elemental geographies, the geohumanities and interdisciplinary scholarship on air and atmosphere. My conceptual proposition, *elemental lures*, highlights the expressions of aerial, atmospheric and meteorological phenomena alongside the cultural legacies and politics of air. Beginning with lures rather than human judgement or the properties of objects enables me to attend to the unruly ways that airy-elemental phenomena affect entities, bodies and materials, and to position human perceptions and interests as emergent from the 'ripples' of the nonhuman and elemental universe.[102] An attention to lures supports a study of the aerosolar arts since aerosolar sculptures transcend the traditional sites and spaces of the art world, drifting, in a very literal sense, in the lower regions of Earth's atmosphere. As I elaborate in the conclusion, the contributions of this work, especially in the interstices of art and geography, may cohere in the *elemental geohumanities*.

In each of the four empirical chapters, I develop creative relations to air and atmosphere, from the studio to stratospheric currents. I do so through different encounters in the aerosolar arts. Chapter 2 considers the elemental lures of shared atmospheres. In particular, the chapter explores the relationship between *Museo Aero Solar* workshops, the atmospheres they conjure and the communities they engage. I borrow the notion of 'interstices' from Philippe Pignarre and Isabelle Stengers to describe the material and affective atmospheres that emerge wherever *Museo Aero Solar* is enacted.[103] These atmospheres connect bodies to each other and to their elemental circumstances. In order for the political potency of the project to be strengthened, however, I argue that a greater attention to the politics of location and the economy of voice is needed. In doing so, I suggest *Museo Aero Solar*'s resurgence in Buenos Aires demonstrates that the project can be productively co-opted for diverse political initiatives. This chapter engages the difficult labour of creating, constructing and sharing atmospheres while showing how these atmospheres tether together different political subjectivities.

Chapter 3 turns to the elemental lures of wind and weather. In this chapter, wind and weather are not only atmospheric phenomena but are also powerful vectors of affect and community politics. Through stories of designing and fabricating aerosolar sculptures in the pedagogical initiatives of *Becoming Aerosolar*, I show that these practices catalyse a phenomenological engagement with air that favours an active questioning of air's materiality and invisible properties.[104] By narrating the 'failure' of several aerosolar sculpture launches, I highlight the perceptions of atmosphere that emerge instead. These include perceptions of the meteorological weather, especially wind speed, cloud cover, aerial convection and heat. They also include perceptions of the weather of law, policy and atmospheric governance.

By amplifying elemental lures of wind and the 'more-than-meteorological' weather, the aerosolar arts invite a phenomenological attention to air that is not about distanced observation but about 'interrogating' air for its simultaneous physical, affective and political qualities.

One outcome of an attention to the lures, rather than the objects of art, is a better grasp of art's movements; these are movements that can extend awareness beyond what is immediately present to experience. In this vein, Chapter 4 examines the elemental lures of transboundary air movements and regulated airspaces through an extended experiment in floating. Following the nomadic flight of two aerosolar sculptures called the *Aerocene Gemini* over one summer day in 2016, this chapter elaborates on the tracking and tracing of aerial entities in commercial airspace. Such methods, carried out by *Aerocene* Community members and radio amateurs, require a suspension of the idea of the object or entity and the ability to read the 'invisible maps' of the atmosphere.[105] Engaging the circuitous route of the *Gemini*, I investigate the relationships between *lures of movement* in the air and dominant infrastructures of aeromobility and aerial surveillance. At the same time, I suggest we bring our attention down to Earth, to map the traces registered and impressed on bodies during these floating journeys. This perspective, I propose, can productively inform an attention to ethics and governance in the *Aerocene* Community.

Chapter 5 explores elemental lures of cloud and sun. Featuring recent launches of a human-piloted aerosolar sculpture – the *D-OAEC Aerocene* – the chapter unpacks the cloud-like images and imaginaries embedded in events of levitation. I show how, with the resources of figuration, metaphor and allegory, or *lures of imagination*, floating bodies become clouds, sculptures become irradiated planets and cloud formations become floating communities. In dialogue with Elizabeth DeLoughrey's discussion of the *heliotrope*, I suggest that practices of levitation conjure uncanny relations between local conditions and planetary processes.[106] More specifically, experiences of levitation in the aerosolar arts telescope between weather and climate in ways that complicate the normative meanings of these terms and enlarge possibilities of imagination and critique. In choreographed levitations and counter-atlases of the sky, artists problematise metaphors of containment and enclosure, question inherited material imaginaries and suggest other relationships between human bodies and planetary atmospheres.

In the book's conclusion, I return to the key gestures of the empirical chapters and suggest what a focus on lures can offer to elemental aesthetics and politics. Through a consideration of works by several artists and artistic collectives, I also suggest how elemental lures may operate in the non-aerial elements. Then I make a call for the elemental geohumanities: a body of thought and practice spanning geography and the arts in which elemental materials, histories and media are the focus of experiment and critique. I suggest that the growth of the elemental geohumanities may contribute to analyses of politics and difference and productively unsettle the inherited material ontologies of geography and related disciplines.

Sensing Art in the Atmosphere is lured by wind, weather, sun, cloud and shared rhythms of breath. Art amplifies these lures, enlarging our capacities to apprehend aerial matters, cultures and politics. Through the currents of *Museo Aero Solar*, *Becoming Aerosolar* and *Aerocene*, this volume investigates the role of art in querying inherited material imaginaries and unsettling normative approaches to elemental air. Whether through the labour of connecting plastic interstices, perceiving from within the current, listening to vast meteorological systems, or imagining clouds and climates, this book senses and experiments with the creative phenomena of air and atmosphere.

* * *

On the second-to-last day we spent at the Salar de Uyuni, an Aerocene sculpture was launched on an 800-m tether and collected video footage. As the camera rose above the sun-drenched lakebed, the tessellated hexagons of the surface below became a faintly discernible texture, while a series of craters suddenly came into view. It was as if we were seeing through the eyes of a probe peering onto the surface of another planet, looking for shapes among the scattering salts.

On the last day, the clouds arrived, and it rained.

Notes

1 For a discussion of Lithium's elemental properties as both pharmacological treatment for manic-depressive or bipolar disorder, and an essential component for the operation of rechargeable batteries, see: J. Lowe, 'I don't believe in God, but I believe in Lithium', *The New York Times*, 25 June 2015. Available at: www.nytimes.com/2015/06/28/magazine/i-dont-believe-in-god-but-i-believe-in-lithium.html

2 For a longer discussion of this excursion and a contextual history of solar, aerostatic flights over salt lakes and sand dunes, see: S. Engelmann, 'Cosmic circuitry', in: A. Oosterman (ed.), *Volume* 47, 2016, pp. 43–48.

3 In defining lures in part via their 'push' and 'pull', I am aware that these terms echo Nigel Thrift's formulation of affect as 'a sense of push in the world' (Thrift, 2004: 64). Although I sympathise with this formulation, in this book I think about lures as the creative, novel, concrete manifestations of what Alfred North Whitehead calls 'propositions' or 'propositional prehensions' (Whitehead, 1978[1929]). These are taken up in more detail in Chapters 3 and 4. Here, I would like to point out similarities and differences between ideas of affect and the propositions of Whitehead. First, like a form of affect, the primary function of a proposition is to induce *feeling*: 'a proposition is entertained when it is admitted into feeling' (Whitehead, 1978[1929]: 188) and it 'exerts a strong "pull"' (Sehgal, 2014: 196). However, unlike affect, defined by theorists (including most famously Brian Massumi (2002)) as the non-conscious, pre-reflective a-signifying 'intensity' that is distinct from conscious, meaningful emotions, propositions can encompass equally 'pulses of emotion' or 'lures for feeling' (Whitehead, 1978[1929]). Indeed as Shaviro (2009) points out, Whitehead uses

'affect', 'feeling' and 'emotion' interchangeably, and it appears he does so in order to work against the anthropocentrism of these concepts. In this way, a focus on lures or propositions is an attention to capacious, nonhuman pulses and currents of feeling, but it is also an attention to the affinity between human experience and broader worlds of feeling. Second, propositions act on the units or 'drops of experience' that Whitehead calls 'actual occasions' (Whitehead, 1978[1929]: 18). Propositions are not (only) found in felt bodily states, material expressions or individual thoughts; rather, they are *presupposed* by bodily states, expressions and thoughts. As Melanie Sehgal explains, 'An explicit thought or phrase is the outcome of the tremendous excitement that propositions induce, if they successfully "lure a feeling"' (Sehgal, 2014: 197). In this book, I follow propositions which are sometimes like, but not equivalent or reducible to, the push and pull of affects.

4 Here I am suggesting that elemental lures, if they take the shape of winds or clouds, are not simply 'natural' phenomena; rather, these winds and clouds also and necessarily heighten our attentions to cultural and political issues and conditions. Nonhuman and synthetic entities can also act as lures in elemental spacetimes. For example, take Derek McCormack's description of a small balloon that he has taken on a train to Glasgow:

> Beyond my immediate attention, the balloon continues, nevertheless, to address me; or, in Alfred North Whitehead's terms, it continues to proposition me. And it does so as a lure for thinking and feeling atmospheres in both a meteorological and an affective sense, offering an invitation to think in the spacetimes between the primacy of process and the intrusive presence of affective materials taking shape as discrete things.

(2014: 609)

5 I borrow the notion of the 'post-electric' from researcher and *Aerocene* colleague Nick Shapiro, who refers to modes of harnessing solar and wind energies that do not rely on batteries and electrical circuits. As he elaborates on aerosolar sculptures, 'Without need to convert solar rays to electricity, these aircraft buck the gridded assumptions of conventional environmental alternatives that are weighty with needs for mined metals and batteries that need to be replaced every six years' (Shapiro, forthcoming).

6 See Y. Bonnefoy, *The lure and the truth of painting: Selected essays on art* (Chicago, IL: The University of Chicago Press, 1995); L. Lippard, *The lure of the local: Senses of place in a multicentered society* (New York: New Press, 1997); M. Bal, *Louise Bourgeois' Spider: The architecture of art-writing* (Chicago, IL: The University of Chicago Press, 2001).

7 For exceptions, see: Monika Bakke's *Going Aerial: Air, Art, Architecture* (Maastricht: Jan van Eyck Academie, 2006) in which she addresses the lures of art in the air, citing Michel Serres description of 'the wild passion of letting yourself be transported by wind, by burning heat and by cold space' (Serres, 1995: 252). Bakke's study is focused on 'individual attempts made by artists, architects and theoreticians to deal with air as information and material carrier, as conductor and catalyst of communication processes' (Bakke, 2006: 15). See also Steven Connor, *The matter of air: Science and art of the ethereal* (London: Reaktion Books, 2010).

8 M. Jackson and M. Fannin, 'Letting geography fall where it may—Aerographies address the elemental', *Environment and Planning-Part D*, 29(3), 2011, pp. 435–444; B. Anderson and J. Wylie, 'On geography and materiality', *Environment and planning A*, 41(2), 2009, pp. 318–335; D.P. McCormack, 'Elemental infrastructures for atmospheric media: On stratospheric variations, value and the commons', *Environment and Planning D: Society and Space*, 35(3), 2017, pp. 418–437;

J.J. Cohen and L. Duckert (eds), *Elemental ecocriticism: Thinking with earth, air, water, and fire* (Minneapolis, MN: University of Minnesota Press, 2015); R. Squire, 'Rock, water, air and fire: Foregrounding the elements in the Gibraltar-Spain dispute', *Environment and Planning D: Society and Space*, 34(3), 2016, pp. 545–563.

9 D. Ferreira da Silva, 'On heat', *Canadian Art*, 2018. Available at: https://canadianart.ca/features/on-heat/

10 P. Adey, 'Air's affinities: Geopolitics, chemical affect and the force of the elemental', *Dialogues in Human Geography*, 5(1), 2015, pp. 54–75; S. Engelmann and D.P. McCormack, 'Elemental aesthetics: On artistic experiments with solar energy', *Annals of the American Association of Geographers*, 108(1), 2018, pp. 241–259.

11 L. Irigaray, *The forgetting of air in Martin Heidegger* (Austin, TX: University of Texas Press, 1999), p. 14.

12 T. Choy, *Ecologies of comparison* (Durham, NC: Duke University Press, 2011). See also N. Calvillo, 'Political airs: From monitoring to attuned sensing air pollution', *Social Studies of Science*, 48(3), 2018, pp. 372–388.

13 H. Bhat, 'Malhar', *Cultural Anthropology* (2019), np. Available at: https://culanth.org/fieldsights/malhar

14 Engelmann and McCormack, 'Elemental aesthetics'.

15 D.P. McCormack, *Atmospheric things: On the allure of elemental envelopment* (Durham, NC: Duke University Press, 2018), p. 6; See also B. Anderson, 'Affective atmospheres', *Emotion, Space and Society*, 2(2), 2009, pp. 77–81; P. Adey, 'Air/atmospheres of the megacity', *Theory, Culture and Society*, 30(7–8), 2013, pp. 291–308.

16 For a critique of the use of atmosphere as a metaphor for the spatiality, flow and morphology of affect, see: B. Verlie, '"Climatic-affective atmospheres": A conceptual tool for affective scholarship in a changing climate', *Emotion, Space and Society*, 33, 2019, online publication ahead of print; for an empirical investigation into affective atmospheres of weather and climate, see: G. Adams-Hutcheson, 'Farming in the troposphere: Drawing together affective atmospheres and elemental geographies', *Social & Cultural Geography*, 20(7), 2019, pp. 1004–1023; examples of works featuring metaphoric senses of atmosphere include T. Edensor, 'Illuminated atmospheres: Anticipating and reproducing the flow of affective experience in Blackpool', *Environment and Planning D: Society and Space*, 30(6), 2012, pp. 1103–1122 and T.F. Sørensen, 'More than a feeling: Towards an archaeology of atmosphere', *Emotion, Space and Society*, 15, 2015, pp. 64–73.

17 McCormack, *Atmospheric things*.

18 I. Stengers, 'Introductory notes on an ecology of practices', *Cultural Studies Review*, 11(1), 2013, pp. 183–196.

19 On the notion of the 'expanded field' and its usefulness in tracing intersections of geography and art, see: H. Hawkins, 'Geography and art. An expanding field: Site, the body and practice', *Progress in Human Geography*, 37(1), 2013, pp. 52–71.

20 My interest in feeling-into and following as an orientation for research is inspired by Isabelle Stengers' extended metaphor of the research scientist as 'tracker' (Stengers, 1997[1989]). In contrast to the 'hunting in a pack' that is practised by scientists who seek to identify, narrow and capture the research problem from every direction, the tracker practises 'empathy', allowing the research site to suggest different and novel directions (Stengers, 1997[1989]: 128; Bhangu et al., 2016).

21 T. Saraceno, S. Engelmann and B. Szerszynski, '*Becoming Aerosolar*: Solar sculptures to cloud cities', in: H. Davis and E. Turpin (eds), *Art in the Anthropocene: Encounters among aesthetics, politics, environments and epistemologies* (London: Open Humanities Press, 2015), p. 59.

22 The Aerocene Float Predictor incorporates real-time information from sixteen-day forecasts of wind speeds at different altitudes. This aerosolar-float trajectory interface is a navigational tool used to plan journeys in the *Aerocene*. Based on a concept by Tomás Saraceno, the *Aerocene* Float Predictor was developed by the *Aerocene* Foundation in collaboration with Lodovica Illari, Glenn Flierl and Bill McKenna from the Department of Earth, Atmospheric and Planetary Sciences at the Massachusetts Institute of Technology (MIT), with further support from Imperial College London, Studio Tomás Saraceno, Radioamateur organisations and the UK High Altitude Society. Atmospheric data is gathered from NOAA's Global Forecast System (GFS), a numerical weather prediction system containing a global computer model and variational analysis run by the US National Weather Service (NWS). The code is open source and available via GitHub: https://github.com/Aerocene/float-predictor

23 In this book, I will often use 'feeling' and 'affect' interchangeably, in line with Alfred North Whitehead's usage. Like Sara Ahmed (2014), I am interested in the complex relationships between bodily sensation, affect and emotion. However, I am also cognisant of arguments by Massumi (2002) as well as several feminist geographers and non-representational theorists who distinguish affect, emotion and feeling as part of an effort to grasp how and why some feelings or affects become recognisable. For a discussion of Whitehead's use of feeling and affect, see: S. Shaviro, *Without criteria* (Cambridge, MA: MIT Press, 2009), p. 46.

24 D. Dixon, H. Hawkins and E. Straughan, 'Wonder-full geomorphology: Sublime aesthetics and the place of art', *Progress in Physical Geography*, 37(2), 2013, p. 232.

25 S. Tooth, S. Viles, H.A. Dickinson, A. Dixon, S.J. Falcini, A. Griffiths and B. Whalley, 'Visualising geomorphology: Improving communication of data and concepts through engagement with the arts', *Earth Surface Processes and Landforms*, 41(12), 2016, pp. 1793–1796.

26 A. McAdie, 'Aërography: The science of the structure of the atmosphere', *Geographical Review*, 1(4), 1916, pp. 266–273.

27 McAdie, 'Aërography', p. 266.

28 Rotch (1911), cited in McAdie, 'Aërography', p. 267.

29 B. Anderson and J. Wylie, 'On geography and materiality', *Environment and Planning A*, 41(2), 2009, p. 319.

30 Anderson and Wylie, 'On geography and materiality', p. 320.

31 A.N. Whitehead, *Process and reality*, Corrected Edition (New York: The Free Press, 1978[1929]).

32 Whitehead, *Process and reality*, p. 185.

33 Whitehead, *Process and reality*, p. 187.

34 Whitehead, *Process and reality*.

35 M. Sehgal, 'Diffractive propositions: Reading Alfred North Whitehead with Donna Haraway and Karen Barad', *Parallax*, 20(3), 2014, p. 197.

36 Sehgal, 'Diffractive propositions', p. 197.

37 McCormack, 'Atmospheric things', p. 609.

38 Anderson and Wylie, 'On geography and materiality', p. 326. It is interesting to note as well that an echo of lures for feeling is again present in the article's final sentence that urges scholars to recognise the 'fallacy of misplaced concreteness': the danger of forgetting the world's processuality (Whitehead (1997[1925]: 58). For Whitehead, it was lures for feeling that partly overcame this fallacy.

39 Anderson and Wylie, 'On geography and materiality'.

40 See, e.g., P. Steinberg and K. Peters, 'Wet ontologies, fluid spaces: Giving depth to volume through oceanic thinking', *Environment and Planning D: Society and Space*, 33(2), 2015, pp. 247–264; R. Squire, 'Immersive terrain: The US Navy, Sealab and Cold War undersea geopolitics', *Area*, 48(3), 2016, pp. 332–338.

41 Jackson and Fannin, 'Letting geography fall where it may', p. 438.

42 Jackson and Fannin, 'Letting geography fall where it may', p. 438.

43 Y. Saito, 'The aesthetics of emptiness: Sky art', *Environment and Planning D: Society and Space*, 29(3), 2011, pp. 499–518.

44 J. Groves, '"The stone in the air": Paul Celan's other terrain', *Environment and Planning D: Society and Space*, 29(3), 2011, pp. 469–484.

45 Jackson and Fannin, 'Letting geography fall where it may'.

46 J.B. Mohaghegh and S. Golestaneh, 'Haunted sound: Nothingness, movement, and the minimalist imagination', *Environment and Planning D: Society and Space*, 29(3), 2011, pp. 485–498.

47 K.R. Olwig, 'All that is landscape is melted into air: The "aerography" of ethereal space', *Environment and Planning D: Society and Space*, 29(3), 2011, pp. 519–532.

48 K. Stewart, 'Atmospheric attunements', *Environment and Planning D: Society and Space*, 29(3), 2011, pp. 445–453.

49 Jackson and Fannin, 'Letting geography fall where it may'.

50 Saito, 'The aesthetics of emptiness', p. 500.

51 In this discussion of the lures of art, I am cognisant of arguments on the 'allure' of objects that are extended in the work of Graham Harman and feature in art history. In Harman's (2011) thought, objects hold something in reserve, and it is this private, withdrawn or 'dark' aspect that is alluring. The concept of 'allure' applies not only to perception by human subjects, but also, 'seeps down even into the heart of inanimate matter' (2011: 302). Shaviro asserts that there is a quality of the aesthetics of the sublime in Harman's concept of allure (2011: 279–290), even though, as Harman (2011) counters, the sublime is an aesthetic theory oriented specifically to the human, while allure is not. Although I will make brief references to object-oriented thought in this book, elemental lures are not informed by ideas of allure in the way Harman and others have developed it. Elemental lures resonate more with Derek McCormack's (2018) treatment of the allure of atmospheres, which is about how the qualities of objects and materials contribute to 'envelopes of experience' in which to think, feel and move differently. In emphasising 'lure' rather than 'allure', I highlight the 'differences' and novelties propagated by atmospheres; however, I do so without hovering, as McCormack (2018) does, between an entity-oriented ontology and atmospheric agency. In addition, for me, working primarily with artistic initiatives and experiments, allure is too narrow a concept and also has too many affinities with art historical ideas of enchantment, judgement and value. Instead, through the abstract and material lures of art, I trace the difference-inducing and imaginatively compelling expressions of air and atmosphere.

52 Adey, 'Air's affinities', p. 54.

53 See G. Bachelard, *Air and dreams: An essay on the imagination of movements* (Dallas, TX: Dallas Institute Publications, Dallas Institute of Humanities and Culture, 1988).

54 Adey, 'Air's affinities', p. 64.

55 Adey, 'Air's affinities', p. 65.

56 Adey, 'Air's affinities', p. 56.

57 J. Thornes, 'Cultural climatology and the representation of sky, atmosphere, weather and climate in selected art works of Constable, Monet and Eliasson', *Geoforum*, 39(2), 2008, p. 574.

58 Thornes, 'Cultural climatology'.

59 Whitehead, *Process and reality*, p. 87.

60 Emphasis mine; G. Born and A. Barry, 'Art-science: From public understanding to public experiment', *Journal of Cultural Economy*, 3(1), 2010, p. 116.

61 Born and Barry, 'Art-science', p. 116.

62 T. Choy, 'Air's substantiations', in: *Ecologies of comparison: An ethnography of endangerment in Hong Kong* (Durham, NC: Duke University Press, 2011).
63 Choy, *Ecologies of comparison*, p. 153.
64 Choy, *Ecologies of comparison*, p. 168.
65 S. Engelmann, 'Toward a poetics of air: Sequencing and surfacing breath', *Transactions of the Institute of British Geographers*, 40(3), 2015, pp. 430–444.
66 Whitehead, *Process and reality*, p. 185.
67 J.J. Cohen and L. Duckert (eds), *Elemental ecocriticism: Thinking with earth, air, water, and fire* (Minneapolis, MN: University of Minnesota Press, 2015), p. 6.
68 S. Mentz, 'Phlogiston', in: J.J. Cohen and L. Duckert (eds), *Elemental ecocriticism: Thinking with earth, air, water, and fire* (Minneapolis, MN: University of Minnesota Press, 2015), pp. 55–76.
69 Mentz, 'Phlogiston'.
70 K. Caplan, *Aerial aftermaths: Wartime from above* (Durham, NC: Duke University Press, 2018).
71 Caplan, *Aerial aftermaths*, p. 58.
72 Caplan, *Aerial aftermaths*, p. 59.
73 Caplan, *Aerial aftermaths*, p. 60.
74 L. Amoore, 'Cloud geographies: Computing, data, sovereignty', *Progress in Human Geography*, 42(1), 2018, pp. 4–24.
75 Amoore, 'Cloud geographies'.
76 Engelmann and McCormack, 'Elemental aesthetics', p. 253.
77 Engelmann and McCormack, 'Elemental aesthetics', p. 255.
78 Engelmann and McCormack, 'Elemental aesthetics', p. 246.
79 C. Sharpe, *In the wake: On blackness and being* (Durham, NC: Duke University Press, 2016).
80 M. Gómez-Barris, *The extractive zone: Social ecologies and decolonial perspectives* (Durham, NC: Duke University Press, 2017).
81 A. Neimanis, *Bodies of water: Posthuman feminist phenomenology* (London: Bloomsbury Publishing, 2017).
82 E.M. DeLoughrey, *Allegories of the Anthropocene* (Durham, NC: Duke University Press, 2019).
83 E.g. Cohen and Duckert, *Elemental ecocriticism*; D. Macaulay, *Elemental philosophy: Earth, air, fire and water as environmental ideas* (Albany, NY: State University of New York Press, 2010); J. Sallis, *Force of imagination: The sense of the elemental* (Bloomington, IN: Indiana University Press, 2000).
84 Y. Furuhata, 'Of dragons and geoengineering: Rethinking elemental media', *Media+Environment* (2019), np. Available at: https://mediaenviron.org/article/10797-of-dragons-and-geoengineering-rethinking-elemental-media
85 Sharpe, *In the wake*.
86 L. Braun, *Breathing race into the machine: The surprising career of the spirometer from plantation to genetics* (Minneapolis, MN: University of Minnesota Press, 2014); S.J. Cervenak, '"Black night is falling": The "airy poetics" of some performance', *TDR/The Drama Review*, 62(1), 2018, pp. 166–169.
87 E. Turpin, 'The Beaufort scale of wind force: This *Land* of forces', in: C. Shaw and E. Turpin (eds), *The work of wind: Air, land, sea* (Vol. 1) (Berlin: K. Verlag, 2018), pp. 9–24.
88 D. Feirreira da Silva, 'On heat', *Canadian Art*, 2018. Available at: https://canadianart.ca/features/on-heat/; S. Langford, '"An atmosphere of certain uncertainty": Knowledge, embodiment and ecology in thick time', lecture and podcast organised by *Vancouver Institute for Social Research*, 25 March 2019. Available at: www.listennotes.com/es/podcasts/visr-vancouver/sasha-langford-an-atmosphere-B2Ug8SYItiK/

89 A. Rich, 'Notes towards a politics of location', in: *Blood, bread and poetry: Se-lected prose, 1979–1985* (London: Virago, 1986), p. 212.
90 Rich, 'Notes toward a politics of location', p. 215.
91 C.T. Mohanty, 'Feminist encounters: Locating the politics of experience', in: L. Nicholson and S. Seidman (eds), *Social postmodernism: Beyond identity politics* (Cambridge, MA: Cambridge University Press, 1995), p. 68.
92 A. Neimanis, 'Feminist subjectivity, watered', *Feminist Review*, 103(1), 2013, p. 25.
93 R. Frankenberg and L. Mani, 'Crosscurrents, crosstalk: Race, "postcolonial-ity" and the politics of location', *Cultural Studies*, 7(2), 1993, pp. 292–310.
94 Gomez-Barris, *The extractive zone*, p. 91. To be clear, Gomez-Barris does not explicitly cite the politics of location and is critical of western feminist materi-alisms. Her approach is to move below the watery surface to access indigenous worldviews and decolonial perspectives emergent from hydro-ecologies of the Americas. Nevertheless, I think her reading of Caycedo's works and her analy-sis of the workings of power within and through dammed and damaged water-sheds bring an important perspective to the discussion of a posthuman politics of location that is enlarged by Neimanis (2017) among others. A thorough en-gagement with the relations between the politics of location, Gomez-Barris' 'submerged perspective' and Neimanis' 'bodies of water' are beyond the scope of this chapter, but they deserve further attention elsewhere.
95 Neimanis, *Bodies of water*, p. 36.
96 Sharpe, *In the wake*, p. 106.
97 Sharpe, *In the wake*, p. 106.
98 Sharpe, *In the wake*, p. 106.
99 Sharpe, *In the wake*, p. 131.
100 A. Neimanis and J.M. Hamilton, 'Weathering', *Feminist Review*, 118(1), 2018, p. 82.
101 J.M. Hamilton and A. Neimanis, 'A field guide for weathering: Embodied tac-tics for collectives of two or more humans', *The Goose*, 17(1), 2018, p. 45.
102 Whitehead, *Process and reality*.
103 P. Pignarre and I. Stengers, *Capitalist sorcery: Breaking the spell* (A. Goffey, trans.) (Basingstoke: Palgrave Macmillan, 2011).
104 Anderson and Wylie, 'On geography and materiality'.
105 I take the figuration of 'invisible maps' from: R. Braidotti, *Nomadic subjects: Embodiment and sexual difference in contemporary feminist theory* (New York: Columbia University Press, 1994).
106 DeLoughrey, *Allegories of the Anthropocene*.

References

Adams-Hutcheson, G. (2019). Farming in the troposphere: Drawing together affec-tive atmospheres and elemental geographies. *Social & Cultural Geography*, 20(7), 1004–1023.
Adey, P. (2015). Air's affinities: Geopolitics, chemical affect and the force of the elemental. *Dialogues in Human Geography*, 5(1), 54–75.
Ahmed, S. (2014). *The cultural politics of emotion*. Abingdon: Routledge.
Amoore, L. (2018). Cloud geographies: Computing, data, sovereignty. *Progress in Human Geography*, 42(1), 4–24.
Anderson, B. (2009). Affective atmospheres. *Emotion, Space and Society*, 2(2), 77–81.
Anderson, B., and Wylie, J. (2009). On geography and materiality. *Environment and Planning A*, 41(2), 318–335.

Bachelard, G. (1988). *Air and dreams: An essay on the imagination of movements.* Dallas, TX: Dallas Institute Publications, Dallas Institute of Humanities and Culture.

Bakke, M. (ed.) (2006). *Going aerial: Air, art, architecture.* Maastricht: Jan van Eyck Academie.

Bal, M. (2001). *Louise Bourgeois' Spider: The architecture of art-writing.* Chicago, IL: The University of Chicago Press.

Bhangu, S., Bishop, A., Engelmann, S., Meulmans, G., Reinhardt, H., and Thibault-Picazo, Y. (2016). Feeling/following: Creative experiments and material play. In *Anthropocene curriculum.* Berlin: Haus der Kulturen der Welt. Available at: www.anthropocene-curriculum.org/pages/root/campus-2014/disciplinarities/feeling-following-creative-experiments-and-material-play/

Bhat, H. (2019). Malhar. In *An anthropogenic table of the elements. Cultural anthropology.* Available at: https://culanth.org/fieldsights/malhar.

Bonnefoy, Y. (1995). *The lure and the truth of painting: Selected essays on art.* Chicago, IL: The University of Chicago Press.

Born, G., and Barry, A. (2010). Art-science: From public understanding to public experiment. *Journal of Cultural Economy*, 3(1), 103–119.

Braidotti, R. (1994). *Nomadic subjects: Embodiment and sexual difference in contemporary feminist theory.* New York: Columbia University Press.

Braun, L. (2014). *Breathing race into the machine: The surprising career of the spirometer from plantation to genetics.* Minneapolis, MN: University of Minnesota Press.

Calvillo, N. (2018). Political airs: From monitoring to attuned sensing air pollution. *Social Studies of Science*, 48(3), 372–388.

Caplan, K. (2018). *Aerial aftermaths: Wartime from above.* Durham, NC: Duke University Press.

Cervenak, S.J. (2018). 'Black night is falling': The 'airy poetics' of some performance. *TDR/The Drama Review*, 62(1), 166–169.

Choy, T.K. (2011). *Ecologies of comparison: An ethnography of endangerment in Hong Kong.* Durham, NC: Duke University Press.

Cohen, J.J., and Duckert, L. (eds) (2015). *Elemental ecocriticism: Thinking with earth, air, water, and fire.* Minneapolis, MN: University of Minnesota Press.

Connor, S. (2010). *The matter of air: Science and art of the ethereal.* London: Reaktion Books.

DeLoughrey, E.M. (2019). *Allegories of the Anthropocene.* Durham, NC: Duke University Press.

Dixon, D.P., Hawkins, H., and Straughan, E.R. (2013). Wonder-full geomorphology: Sublime aesthetics and the place of art. *Progress in Physical Geography*, 37(2), 227–247.

Edensor, T. (2012). Illuminated atmospheres: Anticipating and reproducing the flow of affective experience in Blackpool. *Environment and Planning D: Society and Space*, 30(6), 1103–1122.

Engelmann, S. (2015). Toward a poetics of air: Sequencing and surfacing breath. *Transactions of the Institute of British Geographers*, 40(3), 430–444.

Engelmann, S. (2016). Cosmic circuitry. In A. Oosterman (ed.), *Volume* 47, 43–48.

Engelmann, S., and McCormack, D. (2018). Elemental aesthetics: On artistic experiments with solar energy. *Annals of the American Association of Geographers*, 108(1), 241–259.

Feirreira da Silva, D. (2018). On heat. *Canadian Art.* Available at: https://canadianart.ca/features/on-heat/.

Frankenberg, R., and Mani, L. (1993). Crosscurrents, crosstalk: Race, 'Postcoloniality' and the politics of location. *Cultural Studies*, 7(2), 292–310.

Furuhata, Y. (2019). Of dragons and geoengineering: Rethinking elemental media. *Media+Environment*. Available at: https://mediaenviron.org/article/10797-of-dragons-and-geoengineering-rethinking-elemental-media

Gómez-Barris, M. (2017). *The extractive zone: Social ecologies and decolonial perspectives*. Durham, NC: Duke University Press.

Groves, J. (2011). 'The stone in the air': Paul Celan's other terrain. *Environment and Planning D: Society and Space*, 29(3), 469–484.

Hamilton, J.M., and Neimanis, A. (2018). A field guide for weathering: Embodied tactics for collectives of two or more humans. *The Goose*, 17(1), 45.

Harman, G. (2011). Response to Shaviro. In L.R. Bryant, N. Srnicek and G. Harman (eds), *The speculative turn: Continental materialism and realism* (pp. 291–303). Melbourne: re:press.

Hawkins, H. (2013). Geography and art. An expanding field: Site, the body and practice. *Progress in Human Geography*, 37(1), 52–71.

Irigaray, L. (1999). *The forgetting of air in Martin Heidegger*. Austin, TX: University of Texas Press.

Jackson, M., and Fannin, M. (2011). Letting geography fall where it may – Aerographies address the elemental. *Environment and Planning-Part D*, 29(3), 435.

Langford, S. (2019). 'An atmosphere of certain uncertainty': Knowledge, embodiment and ecology in thick time. Lecture and podcast organised by *Vancouver Institute for Social Research*, 25 March 2019. Available at: www.listennotes.com/es/podcasts/visr-vancouver/sasha-langford-an-atmosphere-B2Ug8SYItiK/

Lippard, L.R. (1997). *The lure of the local: Senses of place in a multicentered society*. New York: New Press.

Lowe, J. (2015). 'I don't believe in God, but I believe in Lithium', *The New York Times*, 25 June. Available at: www.nytimes.com/2015/06/28/magazine/i-dont-believe-in-god-but-i-believe-in-lithium.html

Macaulay, D. (2010). *Elemental philosophy: Earth, air, fire and water as environmental ideas*. Albany, NY: State University of New York Press.

Massumi, B. (2002). *Parables for the virtual: Movement, affect, sensation*. Durham, NC: Duke University Press.

McAdie, A. (1916). Aërography: The science of the structure of the atmosphere. *Geographical Review*, 1(4), 266–273.

McCormack, D.P. (2014). Atmospheric things and circumstantial excursions. *Cultural Geographies*, 21(4), 605–625.

McCormack, D.P. (2017). Elemental infrastructures for atmospheric media: On stratospheric variations, value and the commons. *Environment and Planning D: Society and Space*, 35(3), 418–437.

McCormack, D.P. (2018). *Atmospheric things: On the allure of elemental envelopment*. Durham, NC: Duke University Press.

Mentz, S. (2015). Phlogiston. In J.J. Cohen and L. Duckert (eds), *Elemental ecocriticism: Thinking with earth, air, water and fire* (pp. 55–76). Minneapolis, MN: University of Minnesota Press.

Mohaghegh, J.B., and Golestaneh, S. (2011). Haunted sound: Nothingness, movement, and the minimalist imagination. *Environment and Planning D: Society and Space*, 29(3), 485–498.

Mohanty, C.T. (1995). Feminist encounters: Locating the politics of experience. In L. Nicholson and S. Seidman (eds), *Social postmodernism: Beyond identity politics* (pp. 68–86). Cambridge: Cambridge University Press.

Neimanis, A. (2013). Feminist subjectivity, watered. *Feminist Review*, 103(1), 23–41.

Neimanis, A. (2017). *Bodies of water: Posthuman feminist phenomenology*. London: Bloomsbury Publishing.

Neimanis, A., and Hamilton, J.M. (2018). Weathering. *Feminist Review*, 118(1), 80–84.

Olwig, K.R. (2011). All that is landscape is melted into air: The 'aerography' of ethereal space. *Environment and Planning D: Society and Space*, 29(3), 519–532.

Pignarre, P., and Stengers, I. (2011). *Capitalist sorcery: Breaking the spell* (A. Goffey, trans.). Basingstoke: Palgrave Macmillan.

Rich, A. (1986). Notes towards a politics of location. In *Blood, bread and poetry: Selected prose, 1979–1985* (pp. 210–231). London: Virago.

Saito, Y. (2011). The aesthetics of emptiness: Sky art. *Environment and Planning D: Society and Space*, 29(3), 499–518.

Sallis, J. (2000). *Force of imagination: The sense of the elemental*. Bloomington, IN: Indiana University Press.

Saraceno, T., Engelmann, S., and Szerszynski, B. (2015). *Becoming Aerosolar*: Solar sculptures to cloud cities. In H. Davis and E. Turpin (eds), *Art in the Anthropocene: Encounters among aesthetics, politics, environments and epistemologies* (pp. 57–62). London: Open Humanities Press.

Sehgal, M. (2014). Diffractive propositions: Reading Alfred North Whitehead with Donna Haraway and Karen Barad. *Parallax*, 20(3), 188–201.

Serres, M. (1995). *Angels: A modern myth*. Paris and New York: Flammarion.

Shapiro, N. (forthcoming). Alter-engineered worlds. In Gary Lee Downey and Teun Zuiderent-Jerak (eds), *Making and doing: Activating STS through knowledge expression and travel*. Cambridge, MA: MIT Press.

Sharpe, C. (2016). *In the wake: On blackness and being*. Durham, NC: Duke University Press.

Shaviro, S. (2009). *Without criteria: Kant, Whitehead, Deleuze and aesthetics*. Cambridge, MA: MIT Press.

Shaviro, S. (2011). The actual volcano: Whitehead, Harman, and the problem of relations. In L.R. Bryant, N. Srnicek and G. Harman (eds), *The speculative turn: Continental materialism and realism* (pp. 279–290). Melbourne: re:press.

Sørensen, T. (2015). More than a feeling: Towards an archaeology of atmosphere. *Emotion, Space and Society*, 15, 64–73.

Steinberg, P., and Peters, K. (2015). Wet ontologies, fluid spaces: Giving depth to volume through oceanic thinking. *Environment and Planning D: Society and Space*, 33(2), 247–264.

Stengers, I. (1997[1989]). Is there a woman's science? In *Power and invention: Situation science* (P. Bains, trans.) (pp. 123–129). Minneapolis, MN: University of Minnesota Press.

Stengers, I. (2013). Introductory notes on an ecology of practices. *Cultural Studies Review*, 11(1), 183–196.

Stewart, K. (2011). Atmospheric attunements. *Environment and Planning D: Society and Space*, 29(3), 445–453.

Squire, R. (2016). Rock, water, air and fire: Foregrounding the elements in the Gibraltar-Spain dispute. *Environment and Planning D: Society and Space*, 34(3), 545–563.

Squire, R. (2016b). Immersive terrain: The US Navy, Sealab and Cold War undersea geopolitics. *Area*, 48(3), 332–338.

Thornes, J.E. (2008). Cultural climatology and the representation of sky, atmosphere, weather and climate in selected art works of Constable, Monet and Eliasson. *Geoforum*, 39(2), 570–580.

Thrift, N. (2004). Intensities of feeling: Towards a spatial politics of affect. *Geografiska Annaler: Series B, Human Geography*, 86(1), 57–78.

Tooth, S., Viles, H.A., Dickinson, A., Dixon, S.J., Falcini, A., Griffiths, H.M., and Whalley, B. (2016). Visualising geomorphology: Improving communication of data and concepts through engagement with the arts. *Earth Surface Processes and Landforms*, 41(12), 1793–1796.

Turpin, E. (2018). The Beaufort scale of wind force: This *Land* of forces. In C. Shaw and E. Turpin (eds), *The work of wind: Air, land, sea* (Vol. 1, pp. 9–24). Berlin: K. Verlag.

Verlie, B. (2019). 'Climatic-affective atmospheres': A conceptual tool for affective scholarship in a changing climate. *Emotion, Space and Society*, 33, online publication ahead of print: https://doi.org/10.1016/j.emospa.2019.100623

Whitehead, A.F. (1978[1929]). *Process and reality*, corrected edition. New York: The Free Press.

2   Finding common lures

Museo Aero Solar

I Within the membrane

The Argentine anthropologist Débora Swistun is standing in the middle of an air-filled envelope inside the Palais de Tokyo, a prestigious venue of art and culture near the Champs-Élysées in Paris. Light percolates through the envelope's surface, composed of numerous square and rectangular plastic sheets in many different colours. The membrane flutters slightly, its walls sustained by air circulated by fans through the interior space. It feels constructed, sheltering. A large group of attentive listeners sit cross-legged around Swistun. Speaking of her decades-long research and activism in Villa Inflamable (the 'Flammable Town') on the outskirts of Buenos Aires, a neighbourhood existing in the shadow of one of the largest petrochemical facilities in Argentina, where she was born and raised, Swistun says:

> After some frustrated collective action against pollution [in Flammable] I wanted to understand why it was so difficult to win: at least a resettlement, compensation, cleaning of soils, as I learned [were the outcomes] for several cases in the US of environmental justice.[1]

Bodies adjust and resettle into folds and piles of plastic. Hands play with smooth edges. Faces gaze up and around at the mosaic-like architecture. Turning to her collaborative work with the sociologist Javier Auyero, Swistun continues:

> [In our ethnographic research] we focused on the public dimension of the toxic experiences of Flammable residents, the production of toxic uncertainty, all the actors involved in the labour of confusion about pollution, toxicity, damage on health and the environmental impact in a fenceline community that mainly developed a kind of organic relationship with the petroplants.

The low background hum of ventilating equipment ceases as someone pulls the plug. The plastic surfaces stop their slight fluttering. They hang in suspension. The atmosphere in the envelope grows more still.

> When Javier, my co-author [of the book *Flammable: environmental suffering in an Argentine shantytown*] interviewed some residents, the discourse was somehow homogeneous: 'all of us will die, we are ill, the companies are killing us, and everything is polluted here', they reproduced to him the kind of discourse that [the] media displays about Flammable on the TV they watch and the newspapers they read. When I interviewed them, things looked quite different: doubts, suspicions and denial about pollution, health effects, what the companies do and do not do, if they would be displaced or the companies would be displaced, dominated the conversations.

Swistun pauses here. She takes a breath as she asks us to consider this fact: the means of negotiating toxic atmospheres are contingent on social and collective relations, relations that can be missed by an outsider or a mediated view. Then, she continues:

> We were starting to understand the multiple effects of the (polluted) environmental dimension in urban poverty, a dimension that had been marginal in academic studies about poverty in Latin America. We were bringing to the front the importance of [the] polluted environment in everyday life to understand how social domination works in places like Flammable.

As Swistun speaks, those listening begin to notice the envelope around them getting smaller. The semi-transparent walls are relaxing, slowly giving way to the pressure of outside air. Perhaps unaware, or perhaps knowingly, Swistun presses on, describing the years of activism and community organising during which residents including herself tried to persuade the Supreme Court of Argentina to honour its promise to enforce a 2008 ruling (in Mendoza Beatriz da Silva et al. vs. State of Argentina) to clean up the contaminated basin of the Matanza-Riachuelo River, along which Villa Inflamable is based, and to make reparations for decades of environmental harm.[2] The actions stalled or failed, she explains. Then, she says:

> After ten years of the Supreme Court ruling I still have a question in my mind without a definite answer. And it is as follows: why don't the lives [of people in Flammable] matter as [much as] yours who are here in this room?

This question resonates for several moments. Light still shines through the envelope, but the atmosphere has changed entirely. Slowly a group conversation unfurls, punctuated by increasingly apprehensive stares at the caving surfaces. *Yet no one tries to leave.* As a large fold finally comes gliding down from the vaulted ceiling to the floor, separating half of the group from the other, the conversation stops abruptly. Several tall bodies hold the space open as others squeeze out through a floor-level hole in the now deflated envelope that had, moments earlier, seemed so unshakeable.

Swistun's address inside the membrane of *Museo Aero Solar*, a collectively constructed aerosolar sculpture exhibited in the Palais de Tokyo on the occasion of Tomás Saraceno's Carte Blanche exhibition *On Air*, invites us to address the elemental lures of shared atmospheres. These are atmospheres of being and breathing in common, in which the elemental properties of air inflect the transmission of affect. As Sara Ahmed argues, atmospheres are oriented, and they are 'angled' differently for different bodies.[3] Atmospheric spacetimes have a 'tilt'.[4] Instead of smoothing the flow of affect among bodies in the room or celebrating the qualities of the air/space in the membrane, Swistun's words amplified the angles, the tilts, of the shared atmosphere. The fact that Swistun spoke these words in a symposium at the Palais de Tokyo, while standing inside *Museo Aero Solar*, has a range of important meanings that will be unfolded throughout this chapter. To unpack this moment is to hone an analysis of atmospheric gatherings. It is also to think through an elemental politics of location: to consider how bodies and histories are implicated in local elemental conditions, while weathering forces that extend far beyond them. It also leads us to ask critical questions about the capacity to form communities, artistic or otherwise, that operate within and across disparate political and elemental contexts.

To further engage these issues and questions, this chapter traces part of the history of *Museo Aero Solar*, understood as both a shifting community of practice and a particular type of nomadic, aerosolar sculpture. Over its lifetime, *Museo Aero Solar* crossed from Global North to Global South and back again, forming affiliations with different places, initiatives and struggles. By highlighting some of these 'landings', we can identify *Museo Aero Solar*'s 'interstitial' politics, born of the collective work of fabricating, enveloping and floating with the sun and the air. We can also explore the project's relationships to place and site. Among the aerosolar arts considered in this volume, *Museo Aero Solar* is the oldest, yet this does not mean it is free from problems of collective organisation, nor does its age guarantee its persistence. Indeed, like other manifestations of the aerosolar arts, and as allegorised in the parable that begins this chapter, *Museo Aero Solar* hovers between the poles of continuity and collapse.

By examining the properties of shared atmospheres, the politics of location and the history of *Museo Aero Solar*, this chapter investigates the labour of *finding common lures*. This phrase refers to the process of locating shared investments, interests and desires for the present and future. A common lure acts on multiple bodies, beings and entities. Finding common lures is a crucial dimension of community-based practices within and beyond the aerosolar arts. An initial interpretation (one that this chapter will complicate) might

Figure 2.1 Museo Aero Solar, 2007–ongoing; at ON AIR, carte blanche exhibition to Tomás Saraceno, Palais de Tokyo, Paris. Curated by Rebecca
Lamarche-Vadel.

Note: Initiated by artist Tomás Saraceno in conversation with Alberto Pesavento in 2007,
Museo Aero Solar unfolds in the space formed between human and nonhuman participants in
the simple acts of cooperation and reusing plastic bags, to collectively produce an aerosolar
sculpture. Fostered in more than twenty-one countries to date, *Museo Aero Solar* embodies a vision of pollution-free futures through the growth of self-assembling, geographically
dispersed participatory communities; in this way, the practice can be seen as marking the
beginning of the genealogy of *Aerocene*.

Source: Courtesy *Museo Aero Solar* and *Aerocene* foundation. Photography by Aurèlie Cenno,
2018. Licenced under CC BY-SA 4.0.

suggest that the enactment of *Museo Aero Solar* connects bodies to each other
and to airy-elemental conditions, enlarging new political possibilities. However, as Swistun's address revealed, and as this chapter will show, there are
limits to what is shared atmospherically and what is held in common. For all
of these reasons, addressing the elemental lures of shared atmospheres requires careful attention to the politics of location, to the role of *interstitial*
practices, and to economies of voice in aerial-elemental spaces.

II *Museo Aero Solar*

To engage with the elemental lures of shared atmospheres, we begin by tracing the history of *Museo Aero Solar*. This history reveals the politics and
aesthetics of aerosolar initiatives, and it aids in understanding the simultaneous material and affective dimensions of elemental spacetimes. For purposes of definition, according to the project's webpage, *Museo Aero Solar* is

...[a] flying museum, a solar balloon completely made up of reused plastic bags, with new sections being added each time it travels the world,

thus changing techniques, drawings and shapes, and growing in size every time it sets sail in the air.[5]

However, in the words of Jatun Risba who authored a manifesto for the project, *Museo Aero Solar* is 'more than a balloon' or an 'unusual, mobile art sculpture' because of its emphasis on self-management, self-organisation and self-financing.[6] In another sense, and according to Michela Sacchetto who gave a seminal speech on behalf of the *Museo Aero Solar* community at the Roskilde Festival in Denmark, '*Museo* contributes to an archaeology of the daily life of contemporary society, in working as a conservatory of plastic bags and the memories related to them'.[7] Sacchetto expands: 'it stands as an exercise in the *reduction of boundaries* between everyday people and technical and physical devices'.[8] As I will show, *Museo Aero Solar* reduces boundaries and invents *interstices*.

The first *Museo Aero Solar* sculpture workshop took place in 2007 in the United Arab Emirates, on the occasion of the eighth Sharjah Biennale, where Tomás Saraceno presented the installation *Air-Port-City* (2007) and the guide *59 steps to be on air – by sun power*. In the words of independent curator Murtaza Vali, the eighth Sharjah Biennale was a significant context because 'it encouraged an ecological rather than object-based practice, asking artists to respond specifically to and interact with Sharjah's environment and inhabitants'.[9] Invoking ecological and community-specific practices, Vali echoes arguments of 'new genre public art' that privilege engagement with everyday audiences and the social and environmental issues they face. *Museo Aero Solar* had its first instantiation among artworks that involved, for example, installing functioning solar-powered desalinisation devices in local schools (Marjetica Potrc) or raising the temperature of the museum by two degrees Celsius (Tue Greenfort). Images of the *Museo Aero Solar* workshop show participants, including many women and children, cutting and taping plastic bags in a vaulted hangar.

However, according to Sacchetto and others, it wasn't until Saraceno was invited to a residency at the Isola Arts Centre in Milan that *Museo Aero Solar* achieved a more recognisable form. There, Saraceno worked with Alberto Pesavento as well as hundreds of local volunteers in the Art and Neighbourhood Center in Isola.[10] As part of the exhibition and project *SituazionIsola: A New Urbanism* curated by Marco Biraghi, Maurizio Bortolotti and Bert Theis, the collectively made *Museo Aero Solar* sculpture responded to Milan's Situationist legacy, as well as notions of 'spontaneous aggregation' in urban space.[11] Drawing parallels with other forms of collective making, Sacchetto describes an Isola workshop: 'It was like a sewing class, and the aim was, as it continues to be, to spend some time with other people'.[12] During this period, a number of *tasks* began to differentiate. Some people organised the collection of plastic bags, putting donation boxes and posters in the neighbourhood; others organised children's workshops, while some studied the functional shape for the membrane and mapped the sculpture's meanings. After many trials, the participants in Isola discovered that the best way to inflate the envelope was to channel warm air from urban vents leading from underground transportation systems. Rather poetically, one of the first large-scale *Museo Aero Solar* sculptures was inflated with the 'breath' of an urban neighbourhood. The first launch took place after months of work in the gardens of Isola, which had particular significance as a part of the neighbourhood that had withstood significant pressure from redevelopment forces. Sacchetto elaborates: '*Museo Aero Solar* had its first flight in this very highly charged situation, even if on the day of the flight, there was a really friendly and spontaneous atmosphere'.[13] She makes a crucial point: a neighbourhood can have an atmosphere, one that is sharply oriented or 'highly charged'. Smaller envelopes of experience and experiment can intervene in these conditions, generating shared spaces for communal congregation that are able to occur both because and in spite of atmospheric pressures.

The atmospheres of *Museo Aero Solar* in Isola acted as lures for the future; they profoundly moved bodies, elicited interest and resonated in the imaginations of those involved. According to early community member Till Hergenhahn, the workshops and launches in Isola produced a momentum that continued for the following years. However, the quality of engagement that had occurred at the Isola Art Centre was not necessarily reproducible in other places. *Museo Aero Solar* was invited to the Encuentro Internacional in Medellin, Colombia. On this occasion, Saraceno and Pesavento travelled together as ambassadors of the project. They worked with local groups to prepare the sculpture for one month, at first simply staging the workshop outside and inviting anyone who was interested to join. Although

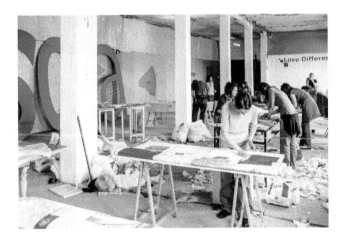

Figure 2.2 Museo Aero Solar, 2007–ongoing; at Milan, Italy, in 2007 with Maurizio Bortolotti, Alberto Pesavento, Tomás Saraceno and the support of Isola Art Center.

my knowledge of this 'landing' is limited, according to Sacchetto, this instantiation of *Museo Aero Solar* engaged the socio-political 'weather' of Medellin, since the aim was 'to develop not only an artistic discussion, but also a political and social discussion about the hard healing process of the city where public life had been deconstructed by years of violence'.[14] It was in Medellin that children began to draw on the plastic canvas, initiating what would become a palimpsest that continued growing in future sculptures. The encounters in Isola and Medellin tested several values of the project: the involvement of local participants in every aspect of fabrication and construction, the feeling of a 'collective property' of actions and a focus on the quality of time spent together. These sites and events also cohered what became known as the *Museo Aero Solar* community: a fluctuating group of volunteers who invested time, funds and energy into developing the project and bringing it to new places. Although Saraceno and Pesavento played important leading roles in these early days, other people including Benedikte Bjerre, Giulia Canta, Janis Elko, Simon Gillard, Till Hergenhahn, Theresa Kampmeier, Daniel Kohl, Dominik Marder, Alice Pintus, Jatun Risba, Tim Rottiers, Michela Sacchetto, Berth Theis and Fani Zguro formed an important *Museo Aero Solar* leadership from 2008 to 2014.[15]

The subsequent 'landings' in Tirana, Kosovo and Ein Hawd, Israel were not as 'pivotal' as those in Medellin and Isola. Sacchetto carefully elaborated that in Tirana and Ein Hawd, *Museo Aero Solar* was presented as a collective endeavour, but in both cases, there was less time to engage with the local community because of poor communication by the art institutions that had organised the encounters. Situations like this reflect art historian Miwon Kwon's argument that community-specific artwork often requires significant mediation from art funding bodies and institutions, which can result in issues of misunderstanding.[16] The repercussions of these events echo in Risba's words, when she says, 'I have serious difficulties in conceiving the art system as being an interesting and appropriate [context] for the Museo'.[17] Nevertheless, these events reveal important dimensions of the atmospheric politics of *Museo Aero Solar*. In Tirana, the day of the *Museo Aero Solar* launch coincided with the day of the declaration of independence of Kosovo from Serbia. Hence, 'by coincidence' *Museo Aero Solar* 'fit very well' in the enthusiastic celebrations.[18] This was due, Sacchetto says, to 'the versatility of Museo in its being a stratification of meanings and in adapting itself to specific context'.[19] In contrast, Hergenhahn states, 'it's very easy to project something onto a balloon'.[20] Thus, *Museo Aero Solar*'s mutability was noted as a feature of its politics from its early days.[21] It's capacity to amplify a joyous atmosphere, Derek McCormack might suggest, is related to its ability to generate an 'atmospheric public' that 'can be intensified, and given volume, through the shape of the envelope'.[22] The degree to which the sculpture produces a public spectacle, or a more nuanced and located gesture, will inform further reflections in this chapter.

An inflection in the project and its politics occurred in 2009, when an invitation from the Walker Arts Centre brought Saraceno and Pesavento to Minneapolis, Minnesota. There, as *Museo Aero Solar* community member Yasmil Raymond documented, they not only organised workshops but also conducted experiments with students of the Aerospace Engineering and Mechanics Department of the University of Minnesota.[23] Together with these students, Saraceno and Pesavento investigated the use of a camera with a GPS tracking system to measure the distance travelled by the sculpture if it flew and they constructed another smaller balloon made of black plastic (perhaps an early precursor of an *Aerocene* sculpture). Saraceno and Pesavento applied for permission from the US Air Force to let *Museo Aero Solar* free fly, but likely due to its size and weight, this request was denied. Nevertheless, the synthesis of knowledge from engineering, applied science and aeronautics is significant because it became a primary feature of later projects like *Becoming Aerosolar* and *Aerocene.*

In the following year, additional workshops and launches took place at the Aeronauten Werkstatt in Frankfurt and Territoria 4 in Prato, Italy. In Prato in 2009, Hergenhahn told me, the organising team faced a crisis when many members lost motivation to launch the sculpture after days of bad weather. On the last possible day, the weather was so fantastic that the sculpture inflated and warmed up quickly, and a child was lifted one metre into the air. The atmosphere of this event has become somewhat legendary: the qualities of the meteorological weather and the combined energies and dispositions of the participants in Prato produced a rare alchemy of the elemental and affective that culminated in a feat of floating.

In the context of these accounts of the atmospheres of workshops and launches, it is important to note that members of the community had long discussed the relationships between the largest and oldest *Museo Aero Solar* sculpture and the construction of younger, lighter sculptures. The organising team agreed that the first, increasingly heavy *Museo Aero Solar* sculpture should become more like a 'cell' that could be divided into multiple entities.[24] They felt that splitting up the single large sculpture would physically represent the nature of *Museo Aero Solar*: 'that it is more like an action, a multiform concept, an instrument to gain other imaginaries'.[25] This is crucial to note: the community placed emphasis on the shareable, processual and imaginative aspects of the project rather than the specific value of one entity or sculpture. Risba even imagined an entire flock of sculptures with different themes, from politics to science to art.[26] The number of sculptures in existence has since fluctuated, but for members of the community, the existence of multiple sculptures is less important than the notion that *Museo Aero Solar* is 'a sharable action that everyone [can] do, start, diffuse'.[27]

An important dimension of *Museo Aero Solar* practices, atmospheres and histories that has not yet been addressed in published writings is the material culture that emerged around the project. As Hergenhahn recounted to me, in the early years, community members would cut out letters from the labels of plastic bags to write notes to each other or to tape humorous phrases onto the membrane. The photographer and community member Janis Elko designed a tote bag with the phrase 'World's Largest Flying Museum', which was reprinted several times. With the small plastic fragments left over from cutting the shopping bags, some members of the community began to make soccer balls. There was a DIY 'starter kit' that consisted of 'a roll of tape, and then an A4 sheet and a drawing going through the tape and then a [specially] folded plastic bag'.[28] There are myriad other stories, from accounts of elderly women who donated single plastic bags covered in carefully written, ornate script, believing these bags would float all the way to heaven reaching loved ones who had passed away, to performances, fashion and design experiments. All of these gestures, from the most subtle and minor to the most public contributed to the shared atmospheres of *Museo Aero Solar* events. Although some elements of this material culture have been absorbed into *Becoming Aerosolar* and *Aerocene*, many others have never been formally collected, studied or exhibited.

Thus far, I have relied on the words and stories of early *Museo Aero Solar* organisers like Sacchetto, Hergenhahn, Risba and Saraceno, archived texts, and publicly available documentation to briefly trace *Museo Aero Solar*'s history. During my time as an ethnographer at Studio Saraceno between early 2014 and late 2016, I also participated in this history through *Museo Aero Solar* workshops in Toulouse, Berlin, Vienna, Paris and Braunschweig. Based on practices learned from these experiences, I led several *Museo Aero Solar* workshops in London, one of which produced a smaller sculpture that was added to the larger membrane exhibited in the Palais de Tokyo in 2018. Many other groups have constructed sculptures, for example in Mississauga, Buenos Aires, Perth, San Francisco and Shanghai. Sometimes the manifestations of *Museo Aero Solar* have taken on other names. In December 2014, during his visit to Lima, Peru upon an invitation from Pablo Suarez of the Red Cross Climate Centre to participate in the Development and Climate Days Conference, Saraceno named the *Museo Aero Solar* sculpture constructed there *Intiñan*, Quechua for, '*the way of the sun*'.[29] The renaming of sculptures suggests that *Museo Aero Solar* may take on specific messages and meanings in particular sites. As Sacchetto and Hergenhahn suggest, this 'versatility' can lend an affective potency but can also complicate the project's relationship to specific places and struggles, as will be expanded further on.

According to Hergenhahn, and at the time of writing, there is a large *Museo Aero Solar* 'cell' at the Aeronauten-Werkstatt in Frankfurt. Another one has been kept in Tomás Saraceno's studio and was exhibited at the Palais de Tokyo, where this chapter began.[30] As one of the oldest *Museo Aero Solar* sculptures in existence, it is composed of sections made in over 20 countries. The logos on the plastic bags vary in languages and meaning. Initially a seemingly secure structure, the walls of this atmospheric entity collapsed in the aftermath of Swistun's injunction: *why don't the lives [of people in Flammable] matter as [much as] yours who are here in this room?*[31]

In addition to 'angling' the shared atmosphere in the envelope, this question invokes an elemental politics of location. Swistun gestured to the historical and bodily exposures of those living in Villa Inflamable, and simultaneously, to the socio-political and institutional frameworks that sustain forms of violence and oppression in Argentina. The name 'Flammable' vividly conveys the precarity of life in the shantytown: it originates from 28 June 1984 event of a fire on the *Perito Moreno*, an oil tanker harboured in a nearby canal. As is carefully expanded in the book *Flammable: Environmental Suffering in an Argentine Shantytown* by Javier Auyero and Débora Swistun, Villa Inflamable is a place where '50 percent of the children tested... have blood levels that exceed the level at which people are declared to be lead poisoned'.[32] According to government-funded epidemiological studies, it is a place where inhabitants are exposed to chromium, benzene and toluene as a result of nearby petrochemical, waste-processing and incinerator facilities.[33] Auyero and Swistun's ethnography employs photographs taken by children of the neighbourhood; many of these images feature trash in lagoons, on streets or near houses. In a series of fieldnotes collected on 8 January 2005, Swistun describes an incident of Villa Inflamable's *weather*:

> I'm still inside. I don't know whether the smell is still out there. I can't smell a thing here. Every time there's this kind of odor we close everything; it is as if we are surrounded by decomposed garbage. After five minutes or so, my mom tells me that the smell is gone. She tries to convince herself by saying: "It might be the weather." "Yes," I reply, "it's Flammable's weather... the stinking smell of the *cinturón ecológico* [the name of the nearby landfill]".[34]

For residents of Villa Inflamable, the waste and petrochemical industries are not only experienced as effects on the climate; they also create local weather. In their ethnographic writing, Auyero and Swistun marshal testimonies that speak to an infinity of such events. Yet, Auyero and Swistun argue, because of contradictory, manipulative and incongruous actions of state, corporate and municipal officials, residents of Villa Inflamable exist in a state of 'toxic uncertainty', trying to maintain daily rhythms as they are repeatedly disrupted by elemental patterns and atmospheres unique to the neighbourhood.[35]

All of this, and much more, was held within the words Swistun spoke in *Museo Aero Solar* in October 2018. Whether *Museo Aero Solar's* subsequent deflation was intentional or accidental, in reflecting on the event since, I think it is significant that it occurred in this way and at that time. It is significant because it evoked a visceral reaction to Swistun's words: the deflation of the sculpture can be understood as a puncture in the fabric of the symposium and the wealthy Parisian venue as Swistun pointed out the historical privilege and inequity on which the conditions of the gathering were predicated. One could also read this event as a demystification of the spectacle of community-centred art. At the same time, the sculpture's deflation allegorised the fact that no matter how different a group of bodies might be, they share a common envelope of breath and atmosphere. The caving envelope of *Museo Aero Solar* reflected the incommensurability of elemental experience while co-implicating an assembly of bodies in atmospheric conditions, indeed physically and affectively *pressing* on bodies as they evacuated the membrane.

To further elaborate on the shared atmospheres of *Museo Aero Solar* with a focus on its relations to local politics, I will narrate several other 'landings' while borrowing from Philippe Pignarre and Isabelle Stengers' notion of 'the interstice'.[36] The term 'interstice', which comes from the Latin *inter-sistere*, meaning 'to stand between', has a physical meaning epitomised in the multiple edges and connected pieces of a *Museo Aero Solar* sculpture, as well as a political and geographical one manifested in the refrain of *Museo Aero Solar* workshops in time and space. It has an atmospheric meaning, too, in the sense that atmosphere has often been understood as that which circulates between and among bodies, and through the pores of bodies, devices and things.[37] I employ interstices as descriptors of *Museo Aero Solar* sculptures and the relations they engender between materials, bodies and landscapes, and as figurations for the atmospheres that emerge in these aerosolar events. Thinking about *Museo Aero Solar* as an interstice enables a partial diagnosis of its political capability: one that emerges in situations where reworking a material to create an envelope of air becomes a way of working with and through differences. At the same time, these interstices must be matched with a politics of location and an attention to the economy of voice if their political potency is to be strengthened rather than vacated.

III Interstitial practices

Museo Aero Solar is, in a literal sense, an object of *interstices*. This aerosolar work is animated by a design, a concept, and a way of practising that is about connection and tension applied to a raw material, that of reused plastic shopping bags. The bags are cut into rectangular shapes and connected using scotch tape, glue or a hot iron. The process and materiality of *Museo Aero Solar* suggests that 'knowing what a material is capable of when a tension is applied to it – recovering its original form, remaining folded, breaking – depends on interstices'.[38] In the case of *Museo Aero Solar*, it is the collective strength and consistency of plastic bags that determines the envelope's capacity to inflate and become airborne. However, the interstices are not only those of bag-to-bag seams but also those of hand-to-hand, body-to-membrane and body-to-body. Another important interstice is the one between *Museo Aero Solar* and the local atmosphere in which it is performed. As I have begun and will continue to show, the interstitial qualities of *Museo Aero Solar* enable it to be reproduced in many places. In this way, the project is shared without the need for enormous budgets, expert skills, expensive materials or specific venues. While scholars like Kwon have critiqued participatory art that travels to 'one place after another', the interstitial qualities of *Museo Aero Solar* belie another logic.[39] In contrast to 'new genre public art' or 'audience-specific' collaborations beginning with (vague) definitions of community and specific social issues, a process that can exacerbate power relations and 'remythify' the artist,[40] with some exceptions, *Museo Aero Solar* is more nebulous and pragmatic in its determination of community involvement. *Museo Aero Solar*'s process of fabrication – the construction of interstices – means that the project's relationship to site, audience and community is not predefined. At the same time, *Museo Aero Solar*'s interstitial qualities can, but do not necessarily, imbue it with politics sensitive to location and difference.

An attention to the materiality of *Museo Aero Solar* leads to further insights on the particular atmospheres of *Museo Aero Solar* workshops. Although research on arts of making or crafting has tended to favour practices like knitting over reuses of plastic, *Museo Aero Solar* is not unique among maker-oriented communities. The last decade has witnessed a proliferation of groups – from Etsy communities to the global network *Precious Plastic* – that employ used plastic as the basic material of making and craft. In many of these other practices, however, used plastic is melted down, crushed or otherwise reduced to a substance for further manipulation, such as in 3D printing machines. *Museo Aero Solar*, in contrast, works with the existing constraints of used plastic bags as a means for producing a sculpture that can levitate from Earth's surface. In this sense, it remains closer to an ethic of repurposing or repairing that scholars like Carr and Gibson suggest is vital for 'connecting' to the environment in an era of volatility, warming and extinction.[41]

For Martyna Marciniak, an architecture student who participated in many of the *Museo Aero Solar* sculpture workshops in Germany in 2014 and 2015, plastic bags are an ideal 'raw' material for entities becoming lighter-than-air.[42] In an interview, she told me that the homogenous units of plastic bags generate exciting possibilities for experimentation with light, adaptive structures, so that these structures are 'effortless'.[43] What happens when participants collect, cut, wash and tape plastic bags together to form a membrane is an intentional modification of the material thresholds of the individual plastic bags. However, as Marciniak suggests, this modification is not one of forcing a change on the material but of actualising a potential that is latent. In the 1999 film *American Beauty*, for example, a plastic bag dances elegantly with the smaller currents and eddies in the air above an urban sidewalk. When connected to form a continuous membrane, however, plastic bags move with larger forces. This scaled-up form of atmospheric mediation has different affective consequences. As James Ash elaborates, 'Material components and thresholds can be reworked, modified or simply broken down, which in turn generates a whole new set of thresholds within which affects can operate'.[44] What are the new thresholds generated by *Museo Aero Solar*?

In the construction of *Museo Aero Solar*, as in that of other aerosolar sculptures, the goal is to create an envelope. The participatory nature of the work, and the fact that the entity is often constructed outside, means the design of the envelope cannot be too complex. Therefore, *Museo Aero Solar* communities have produced three templates named *Tetro*, *Chico* and *Alto*. They show different ways to create a three-dimensional envelope with minimal connecting, measuring and folding. No matter what final form is achieved, however, there is a palpable change that results as the membrane extends. In one particular time-lapse video of this process, a *Museo Aero Solar* membrane-in-progress is laid flat on a grassy field. In the film, we are viewing the entirety of the membrane from an elevated vantage point, such as a nearby roof. Bodies kneel on the membrane's edges, adding rows of plastic rectangles to expand the surface area and holding it down. As it grows in space, the membrane's surface begins to react to the air around it. While participants circle the edges, adding more plastic extensions, ripples run along the lengthening surface, very much like on the surface of a pond. Soon there are waves of even larger amplitude. One has the feeling that the membrane has been enlivened.

Let's imagine that we are kneeling directly on its surface. The slow labour of constructing the membrane is affectively potent. The repetitive actions of measuring, cutting and connecting become vectors for observation and exchange. During the early stages of the workshop, the collaborative cutting of the plastic into the required shapes provokes observations about what brands the bags are labelled with (where have local people been shopping?) the different kinds of bags (what can we tell from different thicknesses, labels and insignia denoting molecular composition?) and the aesthetics of these entities (what are the properties of colour, texture, smell and pattern that make some more appealing than others?). In the process, personal stories emerge about everyday habits, consumption patterns, cultural differences and human-environment relations. Often, however, the slow labour becomes a vector for the exchange of memories, hopes, anxieties, dreams and fears. In this way, and much like other forms of collective craft, the work facilitates intimate and affective offerings.[45] Or, as Ahmed writes, the membrane becomes an object onto which affects and emotions 'press' and 'stick'.[46] To press is to be impressed; thus, the growing membrane also leaves marks and traces on the surfaces of bodies, whether felt in the rawness of knees or in emotional responses to these surfaces and spaces.[47]

There is another change in threshold, atmosphere and affective capacity as the envelope is folded, inflated, and ultimately, launched into the air. One six-minute-long video of the *Museo Aero Solar (Intiñan)* inflation during the Development and Climate Days conference in Lima, Peru, depicts this striking occurrence. The finished envelope is spread on the lawn of an interior courtyard and is inflated with the use of a large fan. The filmmaker circles the structure and frames the three students holding the membrane open. They laugh and talk excitedly. Soon, the object rears up in a large

Figure 2.3 Museo Aero Solar, 2007–ongoing; at Lima, Peru, in 2014 as an Artis-
tic Experimental Performance during 'Development and Climate Days:
Zero poverty. Zero emissions. Within a generation'.
Source: Courtesy *Museo Aero Solar* and *Aerocene* Foundation. Photography by Studio Tomás
Saraceno. Licenced under CC BY-SA 4.0.

mosaic-wave several meters high. A crowd of conference delegates has
formed on the perimeter. People are recording the event with their mobile
phones. Many young people dash around the flapping envelope holding it in
place. It is clear the sculpture is straining with the pull of the air. The filmmaker points the camera inside the mouth of the sculpture, and the noise of
the fan drowns out all other voices. Inside, a space is opening.

Saraceno enters the camera frame, attempting to hold the corner of the
envelope. He kneels near the mouth where there are several students and
young people crowded around, laughing and smiling. There are loud snaps
and ripples of plastic. Then, he orders the fan to be turned off, and he
communicates to those present that they can go inside. One person crawls
through the opening, then another; then Dr. Chris Field, chairman of the
International Panel on Climate Change (IPCC), follows suit. The camera's
view points toward the floor as the person who is filming also enters. Inside,
a handful of people are standing, arms at oblique angles to the floor, palms
up in gestures of disbelief, not trying to repress their smiles as they gaze
around at the multi-coloured dome. Dr. Field is wearing an uncharacteristi-
cally wide grin. Scattered words are heard. Then, there is a ringing exclamation, from somewhere behind the camera:

This is the best thing that has ever happened to politics!

In an interview, Stengers states, 'I think there may be elation, and laughter, when you feel an 'event' and it makes you alive'.[48] Even when watching the video years later, the affective texture of this event in Lima is palpable. *Museo Aero Solar (Intiñan)* succeeded in radically shifting the atmosphere of a temporary assemblage of humans, technical devices and materials, indeed that of a conference. It is perhaps the unexpected aerial autonomy of *Museo Aero Solar (Intiñan)* – the way it begins to move, shape and react – that is responsible for the emotional response. Each time I participate in a *Museo Aero Solar* workshop I am impressed, physically and affectively, by the intensity of feeling during these moments. The spontaneous exclamation – '*This is the best thing that has ever happened to politics!*' – could be brushed off as a carefree hyperbole. However, an attention to the atmospheric and interstitial qualities of *Museo Aero Solar* would take such hyperbole seriously.

Another reading of *Museo Aero Solar (Intiñan)* might postulate that it is the feeling of being-within, or being-enveloped, which constitutes its primary affective and atmospheric element. One could read the intensity of being-enveloped as a manifestation of Sloterdijk's spherical intimacies.[49] To be enveloped, McCormack writes, is to feel the palpable shape of an elemental spacetime and the duality of the envelope as both atmosphere and entity.[50] However, the flight of the sculpture is equally important to consider. Luckily, its capacity to fly was also captured on film. In this second and much shorter film, the camera is angled from outside the courtyard to include the semi-inflated sculpture. *Museo Aero Solar (Intiñan)* is hovering

Figure 2.4 Museo Aero Solar, 2007–ongoing; at Lima, Peru, in 2014 as an Artistic Experimental Performance during 'Development and Climate Days: Zero poverty. Zero emissions. Within a generation'.

above the grass. Then, slowly at first, it begins to lift. When it is a few feet off the ground, there are a few whistles and shouts, and as it rises even higher a scattered, halting applause breaks out. Then, there is a moment, when the sculpture is about 2 or 3 m above the courtyard grass, when all of a sudden, its velocity changes. It switches from a slow, hesitant, upward drift, to an accelerated ascension: a palpable line of flight. As this occurs, there is resounding applause, punctuated with yells, shouts, exclamations, whistles and cheers. The meteorological atmosphere, animated by the photons cascading from the sun, stirs the affective atmospheres of this event. The sculpture quickly reaches the end of its tether, and the point of its tetrahedron shape turns skyward. As the tether holds it fast to the earth, it shimmers like an otherworldly banner or flag.[51]

Outbursts of emotion, excitement, hyperbole, elation and surprise – these were all part of *Museo Aero Solar (Intiñan)*'s inflation, envelopment and flight. As Edensor explores in relation to events of collective illumination, the generation of atmospheres can reterritorialise everyday spaces 'in ways that re-enchant a sense of belonging and shared habitation'.[52] As we know from Sacchetto and others' accounts, affective atmospheres of celebration, awe and enchantment are common in *Museo Aero Solar* performances. In other words, there are atmospheric dimensions of *Museo Aero Solar* that are relatively consistent across time and space. Alongside the physical work of connecting interstices, and the development of interstitial relations among bodies, materials and atmospheres, we can also think of this qualitative consistency as an interstitial politics: an affective politics formed of the capacity to generate a shared atmosphere and a space in common. It is interstitial not only because of its relation to a collectively fabricated envelope but also because it can be reproduced in many places simultaneously. This politics is interstitial, too, because it materialises in an atmosphere that propagates elementally and affectively among and between bodies.

There are several other dimensions to this interstitial politics. For Pignarre and Stengers, a politics of the interstice finds fissures in unilateral or majoritarian thinking. For these authors, political interstices challenge or rupture dominant codes and infrastructures. A *Museo Aero Solar* workshop is an interstice and a political fissure in the sense that it is potentially disruptive of hierarchies of public space, institutional settings (such as the Development and Climate Days Conference) and the habits underpinning single-use plastic products. Relatedly, for Pignarre and Stengers, interstitial politics is about 'being *infected* by the precarious sense of the possible'.[53] As they elaborate in relation to neo-pagan witch gatherings during the World Trade Organisation protests in Seattle, 'all those who are brought together… are in the first place all 'infected' or 'poisoned' equally, although in different modes, and all equally need what none among them is able to produce alone'.[54] If we consider a *Museo Aero Solar* workshop, in which participants labour for days to create an envelope that can fly temporarily with the sun, the word 'infection' is strangely apt. How else could an entity composed of what is often regarded as a waste material demand such commitment? Infection is useful, too, in accounting for the feelings of atmospheric affect and emotion detailed earlier (*This is the best thing that has ever happened to politics!*). Indeed, as phrased by Sacchetto, one purpose of *Museo Aero Solar* is 'to diffuse, *like a virus*, the action of working on the construction of future Museo(s)'.[55]

For others, however, the atmospheres of *Museo Aero Solar* are angled and tilted in particular ways that have largely unspoken consequences. In other words, there are some potential issues with the interstitial politics and the infectious 'spells' of *Museo Aero Solar*. More specifically, there are issues with the atmospheres of *Museo Aero Solar* launches. Jatun Risba writes:

> ...having personally experienced the strong sensorial aspect of a [successful] flight, its powerful 'magic', I am tempted to say that this magic will last, as in the case of any other spell, only until it won't get trivialized or taken for granted.[56]

Risba goes on to suggest that a launch must be used 'tactically' and should be considered in the context of what it produces in a given situation, a given atmosphere. Indeed, as Saraceno and Hergenhahn told me, the launch in Lima produced sensitivities among some *Museo Aero Solar* community members. Some had issues with the project being associated with the Red Cross, based on their previous experiences of working with the Red Cross in Germany. Others were dismayed to see Red Cross stickers on the membrane of *Museo Aero Solar (Intiñan)* as this represented an institutional branding of the project. Pignarre and Stengers highlight this nuance: 'Interstices need to be protected against the posturing of a hope that lends them a role in search of an actor, and constitutes them as the hostages of this role'.[57] On the other hand, in Lima, the project received an exposure to audiences directly connected to climate change policy and to the working groups of the International Panel on Climate Change. As a symbol and physical manifestation of acclimatised architecture and post-fossil imaginaries, *Museo Aero Solar* precipitated what Blanche Verlie calls a 'climatic-affective atmosphere' in Lima.[58] For all of these reasons, however, the infectious 'magic' and potency of *Museo Aero Solar* does not mean its atmospheres are felt in the same way, are uniformly shared or are necessarily benign.

The term 'infection' takes on still other qualities when we consider that *Museo Aero Solar* has been constructed in sites of environmental contamination, where bodies are at a very real risk of becoming ill from air, water and soil, and in places like Isola, where the atmospheric conditions of life are increasingly rendered toxic by municipal forces. The trope of infection may well describe the affective interstices of *Museo Aero Solar*. However, as other scholars have argued via tropes of 'connection', holding on to the darker meanings of these terms challenges us to think more critically about what, beyond affect and emotion, passes through and among bodies, and

Figure 2.5 Museo Aero Solar, 2007–ongoing; at the San Martín University Center (CUSAM), an educational space created by Universidad Nacional de San Martín, with the Criminal Unit No. 48 of the Buenos Aires Penitentiary Service (SPB).
Source: Photography by Carlos Almeida, 2019.

what has already passed through bodies and the environments they inhabit.[59] These, too, are considerations that can inform the interstitial politics of *Museo Aero Solar*. The following sections return to Buenos Aires to further address the relationship between interstitial politics, the politics of location and the economy of voice.

IV Locating interstices

In this chapter and with the help of many interlocutors, I have narrated the history of *Museo Aero Solar*, attending to the elemental spacetimes and atmospheres of its various landings. I have also attempted, through film, archival documents, interviews and personal experience, to present the affective intensity of *Museo Aero Solar* events. I argued that one of *Museo Aero Solar*'s unique features – its status as a 'shareable action' with an affective potency that can be reproduced across time and space – gives it an *interstitial politics*. As McCormack writes, this is a politics in which the 'elemental conditions of the meteorological atmosphere become a distributed matter of concern'.[60] However, the atmospheres of *Museo Aero Solar* are not free of 'tilt', orientation and power. Moreover, together with Swistun, Sacchetto, Risba and others, I suggested that atmospheres of collective work and practice, however powerful, are, on their own, limited in forging connections to particular bodies, sites and histories.

To address the crucial interstices between atmospheres of collective making and local elemental conditions requires another kind of labour: that of engaging with the imaginative and material investments and desires, or *lures*, of those most deeply implicated in place and site. In other words, the shared atmospheres produced in collective workshops like those of *Museo Aero Solar* are influenced in ways both perceptible and imperceptible, by who is contributing, why they are doing so, and what they *press* and *impress* on the membrane and on each other. I use the phrase 'finding common lures' to refer to these exchanges, more or less porous, between local participants, key issues or concerns, and the membranous envelope. In Isola, the atmospheres of making and launching *Museo Aero Solar* were intimately linked to the material and imaginative interests of neighbourhood inhabitants. Thus, common lures emerged in and through shared labour and invention. As many have noted, these atmospheres in Isola generated significant momentum for the project. The same cannot be said for other sites and encounters. That being said, to address common lures is not to create unhelpful distinctions between *Museo Aero Solar*'s landings, nor to romanticise affiliations to place and site, although I recognise a degree of this is inevitable. Instead, I want to attend to the hard work of finding common lures in collective atmospheric practice, and to trace how atmospheres proliferate, engender and lure others into being.

The *Museo Aero Solar* events that have unfolded in Buenos Aires since 2017 are in many ways indicative of the labour of finding common lures. The events did not begin with an invitation from a museum or cultural centre. Rather, they started in December 2017, when a dinner meeting was held in the house of Joaquin Ezcurra. It was attended by Diego Alberti, Maxi Bellman, Mauricio Corbalan, Sebastian Duran, Justo Sanchez Elia, Margarita Ezcurra, Joaquin Ezcurra, Sofia Lemos, Ines Leyba, Daniel Daza Prado, Tomas Saraceno, Martin Saraceno, Debora Swistun, Pablo Suarez and Pio Torrroja. As Swistun told me, after the group had discussed many aerosolar initiatives, she suggested that one way to 'land' *Museo Aero Solar* and *Aerocene* in a local context in Argentina would be to contribute to an awareness raising campaign in Villa Inflamable on the ten-year anniversary of the Supreme Court ruling (in Mendoza Beatriz da Silva et al. vs. State of Argentina) that ordered companies and municipal agencies to clean up the Matanza-Riachuelo River basin. After planning and liaising between Ezcurra, Swistun and Villa Inflamable activist Claudia Espínola, on 8 July, 2018, a *Museo Aero Solar* workshop took place at the Sembrando Juntos Community Centre as part of a day of activities called *Inflamable tiene voz* (*Inflamable has a voice*). Activities during the day included flights of two *Aerocene* sculptures near the petrochemical facility and an emotional reading by Espínola of a text written by Beatriz Mendoza, the activist who led the 2008 legal action. Instead of reiterating the failings of the court ruling, after which very little had been achieved for the contamination of the local environment or the relocation of industry, in her text Mendoza listed different changes that residents of Villa Inflamable had initiated and

witnessed over the prior ten years. These included obtaining judicial recognition, circulating knowledge on collective conditions and honing organisa-
tional capacities. Mendoza emphasised: 'We learned to shout, even on the
grounds of the Supreme Court of Justice of the Nation and even at the risk
of being expelled from that 'sacred' space'.[61] For Mendoza, this constituted
a 'poetics of resistance'.[62] Mendoza expressed the ongoing struggles of Villa
Inflamable with the metaphor that the community was finding its voice. The
specificity of political struggle in Villa Inflamable and the economy of voice
informed subsequent meetings at the Sembrando Juntos Community Centre
where, as Ezcurra reported, discussion focused on other tactics of gaining
media attention as well as improving the monitoring of contamination in
air, water and soil.[63]

The *Museo Aero Solar* activities in Villa Inflamable joined other instantiations of the project in Buenos Aires that were linked to other political
projects. In May 2019, a piece of the *Museo Aero Solar* membrane originally
constructed in Villa Inflamable became the starting point for another sculpture constructed by scientists of Instituto de Biología y Medicina Experimental (IBYME) and Instituto de Investigaciones en Ingeniería Genética
y Biología Molecular (INGEBI) to protest against budget cuts in science.
Patricia Saragueta, the scientist who led this endeavour, acknowledged to
me in a personal communication: 'The socioeconomic situation in VI [Villa
Inflamable] and in CABA (Buenos Aires City) is completely different. In the
scientific community the idea of constructing a balloon with small plastic
bags from their daily shopping was a ridiculous fantasy'.[64] Still, as soon
as the protest got underway, the attitudes of the scientists changed. *Museo Aero Solar* was inflated at the doorstep of Argentina's Secretary of Science. Employing the slogan 'Science is not disposable', the protest attracted
hundreds of scientists, heads of research teams and members of the public.
Saragueta flew the sculpture, which in her words, 'constituted a poetic act'.[65]
In July, Swistun borrowed the sculpture for a *Museo Aero Solar* workshop
inside the Science Cultural Centre (C3) of the Ministry of Science during the
first Latin American Urban Anthropocene school. A couple of months later,
Ezcurra and several others modified and launched the same *Museo Aero Solar* sculpture during a Fridays For Future climate strike. Camille Valenzuela,
Maxi Laina, Mirko Laina, Olivia Laina, Gabi Sorbi and Amalia Boselli and
many others joined the action.

Simultaneously, *Museo Aero Solar* featured in a series of initiatives organised by Carlos Almeida of the Universidad Nacional de San Martín
(UNSAM). Almeida, a trained puppeteer and theatre designer, was already
running an interdisciplinary academic initiative called the *Inflatable Laboratory* when he first learned of *Museo Aero Solar* and *Aerocene*.[66] Together
with several collaborators in dance, theatre and performance studies, he
was inspired to initiate a second collaborative project called *Inflatable
Laboratory – Large Volumes* featuring *Museo Aero Solar* and a new performance work by dancer Camila Almeida called *Momentum*. In order to collect enough plastic bags to create a large-scale *Museo Aero Solar* sculpture,

Almeida sought the support of the Cooperativa Bella Flor de José León Suárez, a women's recycling collective. A group of women from the cooperative worked with members of the university to separate relatively clean plastic bags from solid waste in the recycling plant. The construction of membranes from the recovered bags was carried out by the students of the San Martín University Center (CUSAM) located in the San Martín prison. Additional workshops occurred in local secondary schools and at the University's Institute of Architecture. At each site, sheet-membranes of approximately 50 m^2 were created. The many fragments were then connected during an event in a circus tent on the University Campus, where the sculpture was also inflated for the first time. Almeida brought his own expertise to these events. Videos posted on social media show Almeida using a network of tethers to 'walk' the gigantic *Museo Aero Solar* sculpture around the vaulted hangar like a huge animated creature.

The *Museo Aero Solar* membranes created by these groups bear the traces of their origins. In images from the work of the Cooperativa Bella Flor, women showcase their decorations for the plastic membrane. They have stitched the plastic into intricate star and flower shapes, and painted these with words like *Paz, Amor, Libertad* and *Justicia*. A short video shows the CUSAM participants shaking out a large *Museo Aero Solar* membrane and simultaneously ducking under it, pulling it down around the outer edges of their bodies so that they are huddled inside the plastic bubble. This 'bubble' makes a stark contrast to the high barbed wire fence surrounding their bodies. In another film from a workshop in a theatre school, a dancer moves inside a vast *Museo Aero Solar* membrane, interpreting its folds in her gestures. Through these examples and those mentioned earlier, *Museo Aero Solar* has achieved a particular currency in Buenos Aires. Rather than 'landing' in these spaces, *Museo Aero Solar* has grown up from within them. In this way, *Museo Aero Solar* has facilitated collective actions and experiments that are implicated in place and site. These actions have been *lured* by the shared investments, interests and desires of different groups of people, and they have left their traces on the surfaces of the membrane. One might even claim there is an emergent material culture of *Museo Aero Solar* specific to Buenos Aires, manifested in new uses of plastic bags, different modes of inscribing them and testing ways of 'tactically' animating sculptures other than launching them. In closing, I briefly gesture to the implications of these events for the aesthetics and politics of *Museo Aero Solar*, and suggest what this contributes to the elemental lures of shared atmospheres.

V Linking interstices

In this chapter, I have explored *Museo Aero Solar* in its first ten years of existence, attending to its affective and political interstices. I investigated how atmospheres of making and launch are not necessarily uniformly felt, equally shared or commensurable with a politics of location. What I find significant in the rapidly unfolding events in Buenos Aires is, first, that individuals and communities have begun to 'co-opt' the project for different political aims and struggles. Based on local perceptions of what *Museo Aero Solar* is and can do, the project is being imaginatively instrumented by IBYME researchers, inhabitants of Villa Inflamable and students of CUSAM and UNSAM, among others. Second, it is significant that *Museo Aero Solar* has facilitated a conversation between a community of scientists and the residents of Villa Inflamable. Saragueta wrote to me: 'The other side of *Museo Aero Solar* is Villa Inflamable'.[67] Through the vector of *Museo Aero Solar*, and the efforts of people like Joaquin Ezcurra, Maxi Laina, Tomás Saraceno, Claudio Espínola and Débora Swistun, Saragueta and others plan to join a gathering in Villa Inflamable to 'work with MAS [*Museo Aero Solar*] and Science'.[68] One of the potential directions is a civic science project studying microorganisms in the wastewater of Villa Inflamable, an effort which may partially meet the desires for media attention and environmental monitoring expressed by Villa Inflamable residents in July 2018.

As others have elaborated in depth, there are issues with 'enumerating the environment' in interactions between science and affected communities.[69] The relationship that *Museo Aero Solar* has enabled between Buenos Aires scientists and Villa Inflamable is not free of these issues. Moreover, there are larger power dynamics in the relationships between the aforementioned groups and actors, who have engaged with *Museo Aero Solar* in very different socioeconomic contexts.[70] These issues of power are accompanied by different economies of voice. As Beatriz Mendoza wrote, the residents of Villa Inflamable are fighting to have a voice, while those operating from within the university and other elite spaces are employing established platforms to fight against the removal of key resources. However, there are also ways that these voices are productively entangled. *Museo Aero Solar* is facilitating a cross-city dynamic between the politics of underfunded scientists and the politics of Villa Inflamable. The 'common lures' among the scientists and Villa Inflamable residents are different, but both are 'sewn' and 'glued' to the mobile membrane. This does not show a formless liquidity of meanings and politics, as much as an emergent exchange in which the hopes, demands, desires and gestures of two normally separate communities are lured and tethered together. Instead of absorbing the veneer of existing institutions, events and spaces, as occurred in some of its past landings, *Museo Aero Solar* is becoming a multifarious device for the construction of community politics driven from the interstices, whether of the neighbourhood centre, the scientific laboratory or the one that links them together.

As I signalled before, *Museo Aero Solar*'s interstitial politics is not free of issues. Yet, as the events in Buenos Aires demonstrate, the project can operate as a powerful affective and political interstice both within and between communities. If past work, including my own, presented *Museo Aero Solar* as an apparatus for sensing differences among those who are enveloped,[71] we can go further in proposing that *Museo Aero Solar* enables the sensing of differences between groups, political struggles and elemental conditions.

This insight has implications for the aesthetics and politics of *Museo Aero Solar*. It would mean, as Risba suggested ten years ago, not assuming that the sculpture must be launched in order to become a tactical atmospheric and affective entity. Rather, *Museo Aero Solar* might be performed in ways that diverge from its status as a balloon. Enabling greater flexibility in the performance of *Museo Aero Solar* would offer other modes of linking the project to shared lures and struggles that might not necessarily include moments of inflation and celebration. Different approaches to performing the sculpture will resonate with a wider set of political positionalities. Thinking of *Museo Aero Solar* in these terms enables a different reading of its collapse inside the Palais de Tokyo. As the walls of the weakened membrane folded-in, another spell was cast: a spell of quiet, concentrated solidarity, of organised escape, and of visceral proximity to an elemental envelope that pressed down upon the life inside it.

To apprehend the elemental lures of shared atmospheres is to reconcile their affective potency and elemental materiality with the forms of power and politics suspended within them. This does not mean resisting feelings of excitement and exuberance when they emerge, but it does mean turning a critical eye to what these affective textures produce and what they may elide. *Museo Aero Solar* has provided a compelling case for this analysis, but a similar attention could be given to the atmospheres of other artistic initiatives or making practices. Although the figuration of the interstice is particularly fitting for the aerosolar arts, it has relevance for aesthetic and political refrains more generally, whether conjured by specific social conditions, rituals or public praxis. Much has been written in geography and art history about the challenges to place, site and location under advanced capitalism. An equally extensive literature has addressed shifting definitions of community. An attention to the affective materiality and interstitial politics of shared atmospheres offers another lure for these debates, one that rescues the situation of being and breathing 'in common' while proposing what the limits to this elemental commonality might be.

Notes

1 D.A. Swistun, 'Speech in *Museo Aero Solar*', *Aerocene* symposium, Palais de Tokyo Museum, 26 October 2018 (programmed as part of the exhibition *Tomás Saraceno: On Air*, curated by Rebecca Lamarche-Vadel).

2 In July 2004, a group of residents of the Matanza/Riachuelo basin led by activist Beatriz Mendoza filed a suit before the Supreme Court of Argentina against the national government, the Province of Buenos Aires, the City of Buenos Aires and forty four companies seeking compensation for damages resulting from pollution of the basin, stoppage of contaminating activities and remedy for collective environmental damage. In July 2008, the Court issued a decision in which it required the national government, the Province of Buenos Aires and the City of Buenos Aires to take measures to improve the residents' quality of life, remedy the environmental damage and prevent future damage (ESCR-net, 2019: np).

3 On the 'angling' of atmospheres, Sara Ahmed writes:

So we may walk into the room and 'feel the atmosphere,' but what we may feel depends on the angle of our arrival. Or we might say that the atmosphere is already angled; it is always felt from a specific point.

(2014: np)

In the scene I am describing in these passages, Swistun's words amplified the sense of an 'angle' in the atmosphere. It made participants notice their feeling the atmosphere from a specific point.

4 On the 'tilt' of atmospheres and 'shared' experiences, see: Ahmed, 'Atmospheric Walls'.
5 *Museo Aero Solar*, 'Introducing a new *Museo Aero Solar* website', 2015, np. Available at: https://museoaerosolar.wordpress.com/2015/05/20/introducing-a-new-museo-aero-solar-website/
6 J. Risba, 'Considerations (for) *Museo Aero Solar*', 2009, pp. 1–2 (text for *Museo Aero Solar* Community, shared with author in 2019). See also *Museo Aero Solar*, 2011.
7 Emphasis mine; M. Sacchetto, '*Museo Aero Solar* speech', on occasion of the Roskilde Festival, Denmark, 2011, p. 3.
8 Emphasis mine; Sacchetto, '*Museo Aero Solar*', p. 5.
9 M. Vali, 'Sharjah Biennial 8 still life: Art, ecology and the politics of change', *Bidoun*, 2007, np. Available at: www.bidoun.org/articles/sharjah-biennial-8
10 Isola Arts Centre, 'SituazionIsola: A new urbanism', 2007, np. Available at: http://isolartcenter.undo.net/index_eng.php?p=1131987191&i=1132933796&z=1186673627
11 Isola Arts Centre, 'SituazionIsola', np.
12 Sacchetto, '*Museo Aero Solar*', p. 8.
13 Sacchetto, '*Museo Aero Solar*', p. 8.
14 Sacchetto, '*Museo Aero Solar*', p. 9.
15 This shortlist was provided to me by Hergenhahn. A complete list of *Museo Aero Solar* community members is available at: https://aerocene.org/buildit/; here I want to note the many other people involved in *Museo Aero Solar*, with the caveat that this list is still partial: Tell Andersson, Saga Asgeirsdottir, Jatun Risba, Simo Barbagallo, Hoxha Besart, Benedikte Bjerre, Bob Sleighs, Juan Camilo, Maria Giulia Cantaluppi, Renaud Codron, Marc Colombaioni, Rolf Degel, Vivana Deluca, Pablito El-Drito, Janis Elko, Fani Zguro, Mara Ferreri, James Flaten, Gaia Fugazza, Giovanni Giaretta, Simon Gillard, Till Hergenhahn, Andria Hickey, Juan Camillo Jaramillo, Rasmus Johannsen, Theresa Kampmeier, Daniel Kohl, Oliver Kral, Dominik Mader, Eduardo Ernesto Marengo, Persichina Matteo, Matteo Mascheroni, Natalija Miodragovic, Mohamed Nageh, Mustapha Nageh, Dragusha Njomza, Marco Orlando, Osmani, Sabine Pahl, Eduardo Perez, Alberto Pesavento, Alice Pintus, Cristian Raimondi, Yasmil Raymond, Christiana Rekada, Barrak Reiser, Jacob Remin Sikker, Hoti Rinor, Hannah Rosales, Iuri Rottiers, Tim Rottiers, Matteo Rubbi, Michela Sacchetto, Hugo Santamaria, Tomás Saraceno, Lahu Saranda, Manuel Scano, Saverio Tozzi, Emek Ulusay, Alejandro Uribe, Mauro Vignando and Lionel Wolberger.
16 M. Kwon, *One place after another. Site-specific art and locational identity* (London: MIT Press, 2002).
17 Risba, 'Considerations', p. 2.
18 Sacchetto, '*Museo Aero Solar*'.
19 Sacchetto, '*Museo Aero Solar*', p. 10.
20 T. Hergenhahn, interview with author, 30 October 2019, np.
21 These statements find further echoes in Derek McCormack's words, when he writes that *Museo Aero Solar* 'remobilizes the allure of early balloon launches and their capacity to capture attention' (2018: 210).

22 D.P. McCormack, *Atmospheric things: On the allure of elemental envelopment* (Durham, NC: Duke University Press, 2018), p. 210.
23 Y. Raymond, 'No bag left behind: *Museo Aero Solar*', Walker Art Centre, 2008. Available at: https://walkerart.org/magazine/bag-left-museo-aero-solar
24 Hergenhahn, interview.
25 Sacchetto, '*Museo Aero Solar*', p. 12.
26 Risba, 'Considerations'.
27 Sacchetto, '*Museo Aero Solar*', p. 12.
28 Hergenhahn, interview.
29 P. Suarez, 'Climate risks, art and Red Cross action. Towards a humanitarian role for museums?', *L'internationale*, 2015, np.
30 These sculptural 'cells' of *Museo Aero Solar* are sometimes called 'the Mother' and 'the Daughter'; however, due to the ongoing transformation of the membranes, the sharing of material from younger, newer sculptures, and the constant re-construction and repair, for Hergenhahn, it is no longer useful to use these terms.
31 Swistun, 'Speech', p. 2.
32 J. Auyero and D.A. Swistun, *Flammable: Environmental suffering in an Argentine shantytown* (Oxford: Oxford University Press, 2009), p. 53.
33 Auyero and Swistun, *Flammable*.
34 Auyero and Swistun, *Flammable*, p. 52.
35 Auyero and Swistun, *Flammable*.
36 P. Pignarre and I. Stengers, *Capitalist sorcery: Breaking the spell* (A. Goffey, trans.) (Basingstoke: Palgrave Macmillan, 2011).
37 Ben Anderson, among others, defines affective atmospheres by way of their in-between properties: they exist 'between presence and absence, between subject and object/subject and between the definite and indefinite' (2009: 77). For an account of atmospheres as they infiltrate the pores and holes of the body, see: M. Nieuwenhuis, 'Porous skin: Breathing through the prism of the holey body', *Emotion, Space and Society*, 33, 2019, online publication ahead of print: https://doi.org/10.1016/j.emospa.2019.100595
38 Pignarre and Stengers, *Capitalist sorcery*, p. 111.
39 Kwon, *One place after another*.
40 Kwon, *One place after another*, p. 6.
41 C. Carr and C. Gibson, 'Geographies of making: Rethinking materials and skills for volatile futures', *Progress in Human Geography*, 40(3), 2016, pp. 297–315.
42 M. Marciniak, personal communication and interview, 20 February 2015.
43 Marciniak, personal communication.
44 J. Ash, 'Technology and affect: Towards a theory of inorganically organised objects', *Emotion, Space and Society*, 14, 2015, p. 87.
45 L. Price and H. Hawkins (eds), *Geographies of making, craft and creativity* (Abingdon: Routledge, 2018).
46 S. Ahmed, *The cultural politics of emotion* (Abingdon: Routledge, 2004).
47 Ahmed, *The cultural politics of emotion*.
48 I. Stengers and M. Zournazi, 'A "cosmo-politics"–risk, hope, change', in: M. Zournazi (ed.), *Hope: New philosophies for change* (London: Lawrence and Wishart, 2002), p. 269.
49 P. Sloterdijk, *Spheres volume 1: Bubbles* (W. Hoban, trans.) (Los Angeles, CA: Semiotext(e), 2011).
50 McCormack, *Atmospheric things*.
51 During a *Museo Aero Solar* launch in Newcastle, Australia in December 2017, this accelerated ascension had other consequences. As recounted by Andrew Styan on the *Aerocene* Forum, a *Museo Aero Solar* sculpture had been constructed over six months with children. It was inflated on a warm summer

morning for 30 minutes. Once it was inflated, however, the sculpture began to rise very quickly, jerked to a stop at the end of the tether, and then broke free, to the amazement and awe of grounded participants. A chase team was not able to retrieve the sculpture, and air traffic control had to be notified. Working with *Aerocene* community member Joaquin Ezcurra, Styan calculated that the sculpture had probably reached 16,000 m in altitude and travelled 600 km, likely landing in the ocean off the coast of Brisbane. Stories like these manifest the overwhelming force and power of enveloped air; they have also led to changes in instructions for building and flying *Museo Aero Solar* sculptures.

52 T. Edensor, 'Moonraking in Slaithwaite: Making lanterns, making place', in: Laura Price and Harriet Hawkins (eds), *Geographies of making, craft and creativity* (Abingdon: Routledge, 2018), p.54.

53 Pignarre and Stengers, *Capitalist sorcery*, p. 138.

54 Pignarre and Stengers, *Capitalist sorcery*, p. 138.

55 Sacchetto, '*Museo Aero Solar*', p. 12. Relatedly, in a presentation that Saraceno and Pesavento gave at the Walker Arts Centre, they stated: 'We think there's more inventive potential among a group of people *gathered in a circle* than in any governmental research agency. The *spread of an invention* is more decisive than the invention itself' (Saraceno and Pesavento, 2009, cited in Sacchetto, 2011, p. 13).

56 Risba, 'Considerations', p. 2.

57 Pignarre and Stengers, *Capitalist sorcery*, p. 115.

58 B. Verlie, '"Climatic-affective atmospheres": A conceptual tool for affective scholarship in a changing climate', *Emotion, Space and Society*, 33, 2019, online publication. https://doi.org/10.1016/j.emospa.2019.100623

59 M. Burke, 'Knitting the atmosphere: Creative entanglements with climate change', in: Laura Price and Harriet Hawkins (eds), *Geographies of making, craft and creativity* (Abingdon: Routledge, 2018), pp. 174–189.

60 McCormack, *Atmospheric things*, p. 210.

61 B. Mendoza, 'Declaración Beatriz Mendoza', unpublished address read by Claudia Espínola at 'Inflamable Tiene Voz', Villa Inflamable, Doc Sud, Argentina, 8 July 2018, p. 2.

62 Mendoza, 'Declaración Beatriz Mendoza', p. 2.

63 The *Museo Aero Solar* activities in Villa Inflamable joined a series of other instantiations of the project in Buenos Aires that animated other communities. In April 2019, within the framework of the cultural project Franja del Rio, a large-scale *Museo Aero Solar* sculpture was constructed by volunteers and artists from Rosario, north of Buenos Aires. The group experimented with inflating the sculpture at sunset, using flashlights to illuminate the interior, and returned at dawn for multiple launches on the bank of the Paraná River. Maxi Laina and Joaquin Ezcurra wrote of this event: 'How can we find new ways to reimagine the politics of the atmosphere, challenge national borders and find creative solutions for our common environment?' (Laina and Ezcurra, 2019: np).

64 P. Saragueta, personal communication with author, 23 September 2019, np.

65 Saragueta, personal communication, np.

66 Almeida attended the Urban Anthropocene School organised by Débora Swistun, where he met Joaquin Ezcurra and learned about the method and design of *Museo Aero Solar* (Almeida, 2019).

67 Saragueta, personal communication, np.

68 Saragueta, personal communication, np.

69 N. Shapiro, N. Zakariya and J. Roberts, 'A wary alliance: From enumerating the environment to inviting apprehension', *Engaging Science, Technology, and Society*, 3, 2019, pp. 575–602.

70 These power dynamics are indicated in the fact that, as a western academic researcher, I was able to interact with Ezcurra, Swistun, Almeida and Saragueta,
but I was not able to speak to current inhabitants of Villa Inflamable in order to
include more of their voices in this chapter.
71 S. Engelmann and D.P. McCormack, 'Elemental aesthetics: On artistic experiments with solar energy', *Annals of the American Association of Geographers*,
108(1), 2018, pp. 241–259.

References

Ahmed, S. (2004). *The cultural politics of emotion*. Abingdon: Routledge.
Ahmed, S. (2014). Atmospheric walls. *Feminist Killjoys* [blog]. Available at: https://
feministkilljoys.com/2014/09/15/atmospheric-walls/
Almeida, C. (2019). Personal communication with author, 6 December.
Anderson, B. (2009). Affective atmospheres. *Emotion, Space and Society*, 2(2), 77–81.
Ash, J. (2015). Technology and affect: Towards a theory of inorganically organised
objects. *Emotion, Space and Society*, 14, 84–90.
Auyero, J., and Swistun, D.A. (2009). *Flammable: Environmental suffering in an Ar-
gentine shantytown*. Oxford: Oxford University Press.
Burke, M. (2018). Knitting the atmosphere: Creative entanglements with climate
change. In Laura Price and Harriet Hawkins (eds), *Geographies of making, craft
and creativity* (pp. 174–189). Abingdon: Routledge.
Carr, C., and Gibson, C. (2016). Geographies of making: Rethinking materials and
skills for volatile futures. *Progress in Human Geography*, 40(3), 297–315.
Edensor, T. (2018). Moonraking in Slaithwaite: Making lanterns, making place. In
Laura Price and Harriet Hawkins (eds), *Geographies of making, craft and creativity* (pp. 44–59). Abingdon: Routledge.
Engelmann, S., and McCormack, D.P. (2018). Elemental aesthetics: On artistic experiments with solar energy. *Annals of the American Association of Geographers*,
108(1), 241–259.
ESCR-net (2019). *Mendoza Beatriz Silva et al vs. State of Argentina et al on damages (damages resulting from environmental pollution of Matanza/Riachuelo
river)*. File M. 1569. XL. Available at: www.escr-net.org/caselaw/2011/mendoza-
beatriz-silva-et-al-vs-state-argentina-et-al-damages-damages-resulting
Hergenhahn, T. (2019). Interview with author, 30 October.
Isola Arts Centre (2007). SituazionIsola: A new urbanism. Available at: http://
isolartcenter.undo.net/index_eng.php?p=1131987191&i=1132933796&z=1186673627
Kwon, M. (2002). *One place after another. Site-specific art and locational identity*.
London: MIT Press.
Laina, M., and Ezcurra, J. (2019). Moving with Franja del Río – Rosario free from car-
bon. *Aerocene*. Available at: https://aerocene.org/franja-del-rio-rosario-eng/
Marciniak, M. (2015). Personal communication and interview, 20 February.
McCormack, D.P. (2018). *Atmospheric things: On the allure of elemental envelop-
ment*. Durham, NC: Duke University Press.
Mendoza, B. (2018). Declaración Beatriz Mendoza. Unpublished address read by
Claudia Espínola at 'Inflamable Tiene Voz', Villa Inflamable, Doc Sud, Argen-
tina, 8 July.
Museo Aero Solar. (2011). Landings. Available at: https://museoaerosolar.wordpress.
com/places/

Museo Aero Solar. (2015). Introducing a new *Museo Aero Solar* website. Available at: https://museoaerosolar.wordpress.com/2015/05/20/introducing-a-new-museo-aero-solar-website/

Nieuwenhuis, M. (2019). Porous skin: Breathing through the prism of the holey body. *Emotion, Space and Society*, 33, online publication ahead of print https://doi.org/10.1016/j.emospa.2019.100595

Pignarre, P., and Stengers, I. (2011). *Capitalist sorcery: Breaking the spell* (A. Goffey, trans.). Basingstoke: Palgrave Macmillan.

Price, L., and Hawkins, H. (eds) (2018). *Geographies of making, craft and creativity*. Abingdon: Routledge.

Raymond, Y. (2008). No bag left behind: *Museo Aero Solar*. Available at: https://walkerart.org/magazine/bag-left-museo-aero-solar

Risba, J. (2009). Considerations (for) *Museo Aero Solar* [text for *Museo Aero Solar* Community, shared with author in 2019].

Sacchetto, M. (2011). *Museo Aero Solar* speech. On occasion of the 2011 Roskilde Festival, Denmark [acquired via Studio Tomás Saraceno].

Saragueta, P. (2019). Personal communication with author, 23 September.

Shapiro, N., Zakariya, N., and Roberts, J. (2019). A wary alliance: From enumerating the environment to inviting apprehension. *Engaging Science, Technology, and Society*, 3, 575–602.

Sloterdijk, P. (2011). *Spheres volume 1: Bubbles* (W. Hoban, trans.). Los Angeles, CA: Semiotext(e).

Stengers, I., and Zournazi, M. (2002). A 'cosmo-politics'–risk, hope, change. In M. Zournazi (ed.), *Hope: New philosophies for change* (pp. 244–272). London: Lawrence and Wishart.

Suarez, P. (2015). Climate risks, art and Red Cross action. Towards a humanitarian role for museums? *L'internationale*. Available at: www.internationaleonline.org/research/politics_of_life_and_death/47_climate_risks_art_and_red_cross_action_towards_a_humanitarian_role_for_museums

Swistun, D.A. (2018). Speech in *Museo Aero Solar, Aerocene* symposium, Palais de Tokyo Museum, 26 October 2018 [programmed as part of the exhibition *Tomás Saraceno: On Air*, curated by Rebecca Lamarche-Vadel].

Vali, M. (2007). Sharjah Biennial 8 still life: Art, ecology and the politics of change. *Bidoun*. Available at: www.bidoun.org/articles/sharjah-biennial-8

Verlie, B. (2019). 'Climatic-affective atmospheres': A conceptual tool for affective scholarship in a changing climate. *Emotion, Space and Society*, 33, online publication ahead of print: https://doi.org/10.1016/j.emospa.2019.100623

3   Lures of perception

Becoming Aerosolar

I Failing to launch

The first time I drove to the German countryside with Tomás Saraceno, Adrian Krell, Sven Steudte, Odysseus Klissouras and his son Ikarus in order to launch an aerosolar sculpture, we failed. It was 19 September 2014, and we had left Berlin at four o'clock in the morning to catch the first light of dawn. We had a silver-coated radar reflector, a basic tracker, and all of the will and energy to release a fragile, pneumatic membrane into the atmosphere. During our drive to the launch site, we drank warm maté and speculated about the weather. We discussed every change in the colour of the sky and the arrangement of clouds. We stopped to guess variations in the direction and speed of the wind. Steudte, a radio amateur, probed other dimensions of the weather: updates from Air Traffic Control, activity on the Automatic Position Reporting System (APRS) and the weather radar. Yet, no matter how much we speculated, a thick layer of stratus clouds obscured the sun's rays, and rain threatened. After some hours, we drove back to Berlin where, ironically, the sun came out for the whole afternoon.

On that day, I learned that whether or not it actually flies, an aerosolar sculpture intensifies local perceptions of air, atmosphere and weather. Unlike heavier-than-air craft or helium-filled balloons, aerosolar sculptures depend on several interrelated factors to become airborne: solar irradiance, temperature, convection, humidity, cloud coverage and the reflective albedo of Earth's surface. They do not enter the atmosphere like objects in neutral space; rather, 'they [are] immersed in a kind of force field set up by the currents of the media that surround them'.[1] Equally, an aerosolar sculpture is affected by institutional forces and by political and social 'weather systems', such as weight restrictions, launch site parameters, insurance schemes and air traffic authorisations. For all of these reasons, the aerosolar arts respond to and amplify elemental lures of wind and weather. For practitioners, this is manifested in enhanced perceptions of cloud, wind and sun as well as law, governance and policy.

This chapter considers the shaping, forming and flying of aerosolar sculptures, namely several made by students I instructed together with Tomás Saraceno, Jol Thoms, Adrian Krell, Natalija Miodragovic, Stefano Arrighi,

Daniel Schulz, Thomas Krahn, Ivana Franke, Sven Steudte, Kotryna Slapsin-skaité and Martyna Marciniak, at the Institut für Architekturbezogene Kunst (IAK). Through accounts that largely feature the 'failure' of aerosolar workshops and launches, I foreground the perceptions that take flight instead. In 2014 and 2015, the work of making, testing and launching aerosolar sculptures was called *Becoming Aerosolar*. This term was used in pedagogical practice (the *Becoming Aerosolar* curriculum) and collaboratively written texts,[2] and it became the title of an exhibition by Tomás Saraceno at the 21er Haus in Vienna that included contributions from our students.[3] The brief period of *Becoming Aerosolar* is significant because it bridges the established community initiative of *Museo Aero Solar* and the sphere of technical and public experimentation that is *Aerocene*. It also encapsulates strong pedagogical elements that have informed aerosolar practices ever since.

The experiments of *Becoming Aerosolar* inform my argument for an expanded and political phenomenology of air, key to the apprehension of elemental lures. As explored in the previous chapter, elemental lures bring our focus to the material and breathable as well as affective and political qualities of atmospheric spacetimes that propel activist initiatives and stir the imagination. The development of elemental lures, however, requires further engagement with the phenomenology of air and atmosphere. Perceptions of air, Gaston Bachelard claimed, derive from air's capacity to make us feel lighter or heavier.[4] Yet as Tim Choy, Jerry Zee and many others have argued, air is not reducible to mass or movement; air transmits myriad qualities and variations of temperature, light, shadow, aridity, humidity and toxicity.[5] According to Eva Horn, an aerial phenomenology or 'aesthesis of air' explores air in all of its sensory forms and variable compositions, from briny fog to travelling spores to street-level ozone.[6] The aerial phenomenology developed in this chapter does not treat air as a fragmented constellation of entities, nor does it narrowly approach air and atmosphere in essentialist terms. Rather, I foreground perceptions that emerge through an 'interrogation' or 'questioning' of air's material, affective and political qualities.[7] The aerosolar arts perform this phenomenological attention to air by staging interrogative encounters with air's hydraulics, amplifying situated perceptions of wind and weather and revealing political striations in local airspace. To elaborate further on the phenomenology of air, atmosphere and elemental lures, we consider the hydraulic 'shaping' of aerosolar sculptures, before turning to a more detailed account of *Becoming Aerosolar*.

II A hydraulic art

Through processes of envelopment, the material, affective and political properties of air and atmosphere are made palpable in particular ways. The engineering of envelopes has a long history inflected with advancements in material science, the machines of aerial war and the labour of craftspeople, balloonists and textile workers, many of them women.[8] As Derek McCormack emphasises, the process of envelopment, understood as the generating of a 'difference' in an elemental milieu is 'not an abstract process but one whose variation is felt in the shape and duration of different bodies'.[9] It is a profoundly tactical process enrolling multiple forms of expertise and material practice. The sensing of variations in elemental air, made tangible in the shape of an envelope and inseparable from the participation of human and nonhuman bodies, is key to the apprehension of elemental lures. This is because the sensing of airy variations enlarges perceptions of air's matters, cultures and politics and, in doing so, generates specific orientations, desires and movements. As collectively engineered envelopes, aerosolar sculptures provide important resources for sensing and elaborating these elemental lures.

Before addressing the process of fabrication, it is useful to outline the physics of solar aerostatic entities and briefly consider the demands these entities make on practitioners. Like other kinds of solar balloon, an aerosolar sculpture gains buoyancy by enveloping air warmed by the energy of the sun. When radiant energy meets the surface of an aerostatic membrane, a fraction of all wavelengths is reflected, a fraction is absorbed, and some may be transmitted. The quantity of wavelengths absorbed is related to the colour, thickness and chemical composition of the membrane. As Junius Edwards and Maurice Long of the National Bureau of Standards showed in 1919 in their study of the effects of sunlight on balloons, dark grey, green and black fabrics with rubber coatings reflect only 5–6% of wavelengths and absorb over 90%.[10] These levels of absorption result in significant temperature increases of the balloon in contrast to the ambient air. At such levels of absorption, the balloon comes close to (but is not exactly) what is known in physics as a 'black body': an entity that absorbs all of the electromagnetic radiation that falls on it.[11] In the case of a solar balloon, energy is also continually *lost* (reflected and radiated) to ambient air. This relationship of absorption and emanation, entropy and negentropy, expresses the shifting relations between light energy, surface materials and bodies of air.

Depending on the quantity of energy absorbed by a balloon membrane and transferred to the enveloped air, how can we understand the relations between pressure, temperature and volume? This process is modelled by the ideal gas equation: $PV = NRT$ where *P is the pressure, V is the volume, N is the number of moles of gas, R is the ideal gas constant* and *T is the temperature.* Temperature and volume are inversely related to pressure: with increases in temperature, molecules of air vibrate faster, and volume expands, while pressure drops.[12] Furthermore, with knowledge of the above variables, one can use Archimedes' principle to predict the buoyancy of a solar balloon.[13] For example, Axel Talon and Etienne Lalique calculated that their 22.7 m³ tetrahedral solar balloon constructed of black polyethylene could lift 1.80 kg beyond its own weight when the temperature of the enveloped air was around 30–32°C and ambient air was 0–5°C.[14] In order to increase its capacity to lift, the balloon must either become warmer (absorbing more radiant energy, therefore achieving a larger temperature differential in comparison to ambient air) or the membrane must be expanded so that there is a larger volume of air inside to generate lift.

While such calculations and predictions are valuable, they do not account for air's convection and turbulence. We need other resources to understand the relationship between a solar aerostat and elemental air. The notion that matter flows – that it has vortices, spirals and hydraulic qualities – is one that Michel Serres draws back to the Greek atomist philosopher Lucretius whose long poem *De rerum natura* posits the universality of laminar flow translated into turbulence (the minimum angle of which Serres names the *clinamen*).[15] Lucretius' poem begins with an image of ocean waters troubled by winds. What Serres calls the *tourbillon* – translated into English as *vortex* – denotes the seemingly unruly yet coherent structures that form in fluids.[16] This, for Serres, forms an overlooked basis for contemporary physics: 'Everything flows, turbulence appears, temporarily retains a form, then comes undone or spreads. Physics is entirely projected on the current events of hydraulics in general'.[17] For Serres, a physics adequate to the expressiveness of matter – aerial, fluid or earthy – would reject stable equilibria and re-embrace the vortex.

Making and launching an aerosolar sculpture means attending to the *tourbillon*: the inclinations, declinations, spins and vortices of air unevenly warmed by the sun. Hence, aerial hydraulics must be taken into account from the early stages of the design process. Following Ben Anderson and John Wylie, this approach makes the design and making process highly *interrogative*: it involves an active questioning of the properties of air, its turbulent movements and its thermodynamic potential.[18] Even under relatively still conditions, the membrane of an aerosolar sculpture reacts to the way energy is distributed in the air. The object folds, twists, doubles back on itself, darts, hovers, rolls up, flattens, shoots up into the air and crashes down to Earth again. Designing, fabricating and launching an aerosolar sculpture requires an iterative questioning of the physics of air, interrogating the aerial capacities of a particular membrane and orienting the practices of humans to apprehend these relations.

Such interrogative practices have a variety of political and perceptual consequences. One of these consequences is recognition of perception as a collaborative achievement of the body and the medium of air. Air is not, as many have assumed, an 'outside', an 'object' or a 'void' distanced from the body. Instead, like a solar aerostat, the perceiving (human) body is forming of, and formed by, air and atmosphere. Thus, the attention to the physics and construction of aerosolar sculptures is potentially productive of an understanding about many other kinds of aerial bodies. This includes an awareness of the relationships between bodies, atmospheres and the weather. Although aerosolar sculptures and human bodies are most exposed to the meteorological weather in attempts to launch, perceptions of the weather emerge long before the act of releasing a sculpture into the atmosphere. In particular, the student-led construction of a series of aerosolar sculptures is a site of interrogation of wind, weather and the 'living air'.

Figure 3.1 Prototyping Aerosolar Sculpture Designs, 2014; workshop in the graduate seminar of Tomás Saraceno, Sasha Engelmann and Jol Thoms at the Institut für Architekturbezogene Kunst, Technical University of Braunschweig, Germany.

Source: Photography by Sasha Engelmann, 2014.

III Making

I addressed the making of the aerosolar sculpture *Museo Aero Solar*, with its particular constraints in materials and techniques, in the previous chapter. The process of making changes when some of these constraints have been removed, when there are a wider variety of tools available and when the operative questions have changed. While the animating concerns of *Museo Aero Solar* revolved around repurposing of materials and the construction of aerosolar entities in decentralised and self-organised modes, *Becoming Aerosolar* focused primarily on the physics of aerosolar sculptures, their optimal shapes and capacities to float into airspace. In addition, the pedagogical spaces that hosted *Becoming Aerosolar* fostered different elemental and affective atmospheres. These were largely enclosed, interior atmospheres of shared focus that mediated the relay of ideas, bodily states, experiences and affects between teachers, experts and students.

One aerosolar workshop in the *Becoming Aerosolar* curriculum occurred in the framework of a graduate seminar taught by Tomás Saraceno, Natalija Miodragovic, Jol Thoms and myself, at IAK in the winter semester of 2014. We worked with 12 masters' students to 'imagin[e] alternative futures in which society is truly 'solar powered,' not [by] using solar panels or batteries – but by harnessing the energy already circulating in air'.[19] We investigated solar-elemental architectures and experimented with the movement of energy through fluid matter. Each week we also discussed topics in atmospheric science and the aesthetics and politics of air. The aerosolar sculpture workshop drew on these discussions and extended key ideas in the course.

Before the workshop, each of the students proposed one aerosolar sculpture design, shared their ideas and formed small groups. They were provided with reused plastic shopping bags, several one-metre-wide roles of PVC plastic (black and transparent), some reflective mylar, as well as scissors, tape, pencils, paper and razor cutters. The workshop included a mix of various specialists. Adrian Krell and Tomás Saraceno offered advice on the architecture of the designs and their viability. Jannik Heusinger, a climatology PhD student at the Technical University of Braunschweig, brought an infrared camera and offered his advice on the surface albedo, interior convection and material absorption. Sven Steudte, a radio amateur, telecommunications and tracking specialist, demonstrated the Arduino micro-controller he had built for communication with airborne entities. Based on the lift calculation of 80 grams per cubic meter of air with a temperature difference of 20°C, the students quickly realised that the more complex models would be either too difficult to construct or would require too much material (and weight) with respect to volume and lift. This was the first major lesson: given the relative difficulty of generating lift by trapping solar energy in an enclosed volume of air, the aerosolar sculptures had to have large volume-to-surface-area ratios.

After some discussion, the students decided to construct several shapes of envelopment: a torus, a Klein bottle, a long cylinder, a diamond, a bi-pyramidal tetrahedron and a sphere. Once the dimensions were decided, the physical work began. In the five-story ex-concrete-factory building that is the IAK, student teams found space to work on open landings, in upper-story classrooms and throughout the large vaulted space at the entrance to the building. In their zones of activity, students and teachers spread out sheets of PVC plastic, cut and taped the surfaces together into wide, flat sheets. They kneeled hour after hour on the concrete floor, shoeless and in pairs, one person holding the tape while the other pulled it across the meeting point of the two surfaces, watching for tears or wrinkles. In contrast to *Museo Aero Solar* sculpture workshops, this was more precise work, informed by architectural modelling software. Much time was spent smoothing the membranes, folding surfaces over each other again and cutting excess material. As night fell, the IAK building, surrounded by a large forest and situated many kilometres from the main university campus, became a glowing hive of activity in the dark.

The careful fabrication process was the source of aches and soreness, bodily impressions of kneeling, cutting and taping, but also, as in *Museo Aero Solar* workshops, it nurtured discussion and affective exchange. Perhaps because of the goal-oriented nature of the work (the students anticipated launching the sculptures at sunrise), and the isolation of the labour away from the university and the city, the atmospheres of making were quiet and focused. Voices were lowered, and furrowed brows marked the faces of participants. Conversations played out in whispers, and tension passed from body to body, as hands touched other hands. These atmospheres changed when a series of tests occurred. Testing an aerosolar sculpture involves opening its 'mouth' toward a fan or ventilator. Sometimes an additional tube has to be fashioned in order to channel air into the inner space. Testing injects air into the sculpture – the hydraulics of the airy medium – to show if the sculpture is strong, flexible and able to float. It is also a test of the shape of the aerosolar body and its ability to hold a form in the turbulent *vortex*: the wind.

A structure shaped like a Klein bottle was inflated around 10pm. As air was funnelled into the envelope, ripples began moving and creasing, undulating in the plastic membrane. The material was snapping back and forth as it was pulled and pushed by the currents of air. In this way, the formerly lifeless membrane was interrogated for its capacity to hold moving air.[20] Quickly, it became apparent that the narrow 'tail' of the Klein-bottle sculpture would not allow easy circulation of air and energy, and so this sculpture was abandoned. The second sculpture was tested close to midnight. It was a multicoloured tetrahedron made of many reused plastic bags reminiscent of *Museo Aero Solar*, and it was fragile due to its many taped seams. As air was channelled inside it, small rips and tears opened. Students rushed around the sculpture brandishing roles of tape. Once the sculpture was stabilised, the process of interrogation suggested that the membrane could indeed hold shape and move in the wind. Nevertheless, it was *not* optimised for extremely windy conditions.

At this point, unexpectedly, one student moved the fan to one side and crawled through the mouth. Several other students followed. Before long, everyone present at the workshop was inside the membrane, talking excitedly, and soon, dancing. Energy was transmitted between bodies of different kinds. Fluorescent light shone through the pale colours and shiny, taped seams into the shimmering interior. The tediousness of the labour was temporarily forgotten as a new social atmosphere came into formation and the creators of the membrane became its inhabitants.

Since the tetrahedron was inflated without bursting and was large enough based on estimates from the lift equation, this design was deemed viable to launch. Still, the number of seams and edges raised a concern that the membrane was too heavy. This was a point of debate:

> Adrian Krell: The joined plastic bags are more heavy than thought. At the moment for example [one] form that we wanted to do… is not flying, because it's too heavy. And for you [pointing to one student] I didn't do the calculation but we have to think it's not twenty or thirty grams per square meter [of plastic], but it's more like seventy grams per square meter…
>
> Tomás Saraceno: It's not so easy you cannot fly everyday… and Adrian and Sven are really keen on trying to do it. At the same time, of all of the designs we have so far, it seems none of them are going to fly so far. They are all too small. This is what Adrian's, and what my experience is, none of them will, even if we have sun tomorrow, nothing will happen…[21]

Krell and Saraceno made two crucial points: that the lift calculations could be skewed due to the weight of the connecting edges (or 'interstices') and that the sculptures were still too small in volume. Following a brief discussion, it was decided that the most viable sculpture was a simple long black cylinder with the least number of edges and the largest volume. In the process, participants explicitly recognised the probability of failure. Although the entirety of the workshop was focused on proposing, testing and adapting of aerial forms, the ensuing conversation further evidenced an interrogation of the hydraulic matters of air and the physical-atmospheric properties of the sculptures. In an 'interrogative mode', 'materialities and liquidities... [offer] imperatives to action which guide, imply, and ordain corporeal sensibilities'.[22] In this workshop, the testing of the sculptures generated specific imperatives to action: the need to use less tape for edges, to expand the sculptures' volumes and to better stabilise the designs. These actions involved particular corporeal sensibilities that affected posture and movement. They also had political and affective consequences, as students took leading roles in the adapting of sculptures (which often resisted these efforts), debated the qualities of aerosolar forms with Saraceno, Krell and other experts (during which the sculptures were assigned individual names), and made collective decisions about process and strategy.

In addition to the imperatives of the sculptures and their relations to the physics of moving air, another zone of questioning involved the weather. It was established that the sun would probably come out between 7am and 9am, although the expected cloud cover changed frequently. The wind was predicted to be gusty and intermittent, dangerous for fragile aerosolar forms. Solar irradiance was calculated based on the time of year and the angle of the sun through the atmosphere at 8am. From this point in the workshop, a relationship to the weather of the launch was established and worked its way into bodies, imaginations and questions. Hovering behind hopeful comments, in furtive acts of checking the forecast, and the making of half-hearted jokes, the weather of the next morning began 'happening' or 'weathering' at IAK close to midnight.

An hour later, the large black cylinder was ready for testing. The sculpture was spread out on the floor and air was blown into the tube. Soon, students began to enter this sculpture as they had done the tetrahedron, checking for tears and feeling the surface. Surprisingly, rather than a uniform black surface, threads of lighter and darker consistency were illuminated, as well as shades of yellow and blue. These illuminated qualities of the interior atmosphere, which had an effect not unlike that of being inside a textured, opalescent skin, generated what Tim Edensor might call affective 'flows' among group members.[23] The experience of being inside was affectively contagious: chants of '*Sun! Sun! Sun!*' reverberated. We ran our hands over the membrane that had, moments before, lain idle on the concrete.

In addressing the stages of this *Becoming Aerosolar* workshop, the material and sensual qualities of atmospheres cannot be separated from their affective and political qualities. Breathing the cold air of the IAK, laden with pollens

Figure 3.2 Becoming Aerosolar, 2014; workshop in the framework of the *Becoming Aerosolar* graduate seminar 2014/15 at Institut für Architekturbezogene Kunst, Technical University of Braunschweig, Germany.

Note: Together with Tomás Saraceno, Sasha Engelmann, Natalija Miodragovic and Jol Thoms, with support from Philip Dreyer, Jannik Heusinger, Adrian Krell, Martyna Marciniak, Matthias Pelli and Kotryna Slapsinskaite, and with the students Caro Brüggebusch, Macarena Cerda, Wanda Gavrilescu, Ariana Hernandez, Guanxi Huang, Xinyu Hou, Max Lingke, Lotte De Lisle, Murial Martins, Nicole Sandt, Camila Ocampo Selbach, Pia Audrey Schlue, Mengjing Shi and Sara Zorlu.

Source: Courtesy *Aerocene* foundation. Photography by Studio Tomás Saraceno, 2014. Licenced under CC BY-SA 4.0.

and particles emanating from the adjacent forest, participants sketched, calculated, predicted, modified, kneeled, cut, taped, smoothed and repaired a series of plastic surfaces. They ran their hands over half-melted PVC and the edges of reused plastic bags, many of which had been touched by other hands in their previous lives. They also inhaled the volatile organic compounds given off by these 'raw' materials, reminiscent of petrol.[24] These molecular clouds, like those of fog described by Elizabeth Povinelli, haunted the workshop with the latencies of the chemical industries.[25] At the same time, they influenced its perceptual and affective economies to the extent that, months later, these odours would be remembered as elements of the workshop. The students cleaned, folded and stacked the sculptures for the trip to the launch site in the morning. As he prepared to catch a train back to Berlin, Bronislaw Szerszynski summarised what had occurred:

> The thing I will take away from today which I will never forget, is the amount of care, and preparation, that you all did, to create these membranes, these dwelling places for air, the air that isn't even in it yet... seeing all your hands clutching the seams and taping them up, and lifting them up in the air, all this kind of care that had to take place in order

to prepare these membranes for the living air, that I will never forget, it was wonderful.[26]

As Szerszynski poetically noted, the hours of preparation anticipated and interrogated the 'living air' – the wind and weather – that would fill the sculptures at dawn. In other words, the labour, the iterative testing of forms and the concentrated atmospheres of the workshop were motivated in part by an anticipation of wind and weather that hadn't yet entered and immersed the sculptures. Yet, in another sense, through processes of testing and adapting, the sculptures were immersed in these currents of 'living air' long before they reached the launch site. The bodies of practitioners, too, were immersed in the 'living air', as they danced inside the envelopes, calculated lift or pondered the movements of high- and low-pressure systems. Szerszynski's words gesture to the elemental lures of wind and weather: the forms of attraction and interest in aeolian vectors and atmospheres that stir the imagination and are felt viscerally in the labouring body. These elemental lures did not emerge spontaneously but were intimately linked to the process of fabricating envelopes that could fly with the sun and the turbulence of air. In this process, many bodies, human and nonhuman, moved into new constellations and were oriented differently to air and atmosphere. The following section narrates the launch of these sculptures to extend these insights into an aerial phenomenology, or 'aesthesis of air'.[27]

Figure 3.3 Aerosolar Sculpture Launch, 2014; part of the *Becoming Solar* graduate seminar of Tomás Saraceno, Sasha Engelmann and Jol Thoms at the Institut für Architekturbezogene Kunst, Technical University of Braunschweig, Germany.

Note: With Caro Brüggebusch, Macarena Cerda, Wanda Gavrilescu, Ariana Hernandez, Guanxi Huang, Xinyu Hou, Max Lingke, Lotte De Lisle, Murial Martins, Nicole Sandt, Camila Ocampo Selbach, Pia Audrey Schlue, Mengjing Shi and Sara Zorlu at the Institut für Architekturbezogene Kunst, Technical University of Braunschweig, Germany.

Source: Photography by Sasha Engelmann, 2014.

IV Launching

We woke at 5am the next morning, piled into a few cars and chased the rising sun down ghostly highway roads. At 7am, the sculptures were unfolded on a hilltop in Wolfenbüttel, the location that Steudte and Krell had arranged in agreement with local Air Traffic Control. The array of sculptures (two large and five small) looked like colourful sea creatures sprawled on the dewy grass, beginning to breathe. It was an icy morning; teachers and students hopped in place or huddled together. As the sun rose, piercing through gaps in the low stratus and cumulous clouds on the horizon, the mouths of the sculptures were alternately held open to the wind and then closed so that the air inside could warm. As they inflated, and their shapes gained definition, an increasing feeling of tension passed through the group: bodies sensed each other within and through the turbulent air. Or, as Kathleen Stewart writes, 'senses sharpen[ed] on the surfaces of things taking form'.[28]

Although the sunlight, filtered by clouds, was relatively weak, the air was so cold that the black membrane of the large cylindrical sculpture warmed faster than the ambient air. Holding a hand to the black membrane, one could feel the difference in temperature. A slow and steady exchange was occurring between photons, the electron energy-states of chlorine atoms in the PVC membrane, and the molecules of air inside it: entropy was increasing. Although it was held to the ground in several places, the wind caused the sculpture's membrane to snap and roll around. These movements were far more erratic than what we had witnessed during the test inflation the evening before. In Whitehead's terms, the 'creativity' of the wind – its novelty and complexity – far surpassed the air blown by the fans during the workshop.[29] Then, a series of events unfolded quite rapidly.

I found myself near the long cylindrical sculpture, now nicknamed 'Frankie'. Someone handed me the radar reflector and the GPS tracking device, attached with tethers to its mouth. Two students were holding the mouth shut so that the air inside would warm and expand. The sculpture was making a lot of noise, flapping in the wind and undulating back and forth. Steudte was speaking in German to Air Traffic Control, listening for an official 'OK'. Krell and Saraceno were running around spotting rips and yelling instructions. We waited. I could feel the membrane pulling…

Finally, Steudte got clearance from Air Traffic Control. Saraceno and Krell motioned to release the membrane. We retreated to let it rise. It levitated and began to move. As it floated over the grass, Saraceno, Krell and I jogged along with it. It was moving quickly, lifting upward, expanding. I could feel the pull of the sculpture in my whole body. It was several times my height. The mouth of the balloon yawned. I looked upward, directly inside. The colours were golden, grey, brown and blue. It moved, creature-like, with complicated folds and twists, amplifying eddies and vortices in the air, lines waving and bending through the medium and membrane, joining and spinning away.

As my gloved hand rose to let go of the ropes, there was a slashing sound: a rip opened on the back, letting the pale blue sky show through for an instant before closing as the sculpture became a kite, gliding down to the grass at the bottom of the hill. The launch failed. In fact, we did not succeed in launching any of the sculptures that day, due to the strength of the wind on that hilltop, the weak sunlight through shielding clouds, and the weight and relatively small volume of the sculptures. Nevertheless, the experience of the sculpture lifting off the ground and hovering over the grass, going at the exact speed of the wind, while moving with the micro-eddies of the medium, was registered acutely in my body. In particular, my awareness of local weather conditions (including cloud cover, solar irradiance and wind) was intensified in perceptions of the multiple directions of air movement, the work of solar energy in levitating the sculpture, the swirls, vortices and patterns of the wind, and the speed of the ascent. Matching my gait to that of the sculpture was a challenge in sensing thermodynamic potential: the transactions between electrons, photons and an elemental body of air within an ultra-light structure. In this way, I later reflected, the sculpture and I entered the flow of the 'weathering world' as it entered us.[30] In the context of the collective attempt to launch, the sculpture also enlarged my experience of wind and weather as simultaneously affective and political.

No matter how powerful my experience was at that moment, however, I find it unhelpful to privilege the perceptions of one person holding the sculpture's tethers because so many bodies were also entrained and lured to the wind and weather at that moment. Everyone present that morning was involved in the readying of the sculptures and perceived their responses to air currents, cloud, humidity, heat and light. To perceive in this way is not to practise a form of distanced observation. As Tim Ingold writes:

> …there are no objects of perception. They are rather what we perceive with. In short, to perceive the environment is not to look back on the things to be found in it, or to discern their congealed shapes and layouts, but to join with them in the material flows and movements contributing to their – and our – ongoing formation.[31]

In this formulation, perception is a perceiving-in and perceiving-with an environmental milieu already present in our bodily formation, our being. This is a phenomenology of perception that moves away from observations of fixed shape and substance toward recognition of the medial and phenomenal bases for perception itself. If as Eva Horn writes, air is the most medial of substances, a phenomenology of air is one that foregrounds 'the perception of wind, chill, moisture, breath, the smells and texture of air, a world that is both ever-shifting and changing, in flux, but also situated, grounded and grounding, repetitive'.[32] Rather than privileging an individual experience as constitutive of enhanced perception of wind and weather, the events presented here have more purchase for grasping 'situated', 'ever-shifting' and 'repetitive' perceptions of air that pre-exist the clouded hilltop, that provoke corporeal capacities and that viscerally manifest in bodies and imaginative faculties. As Szerszynski observed, these orientations to air were there in the 'dwellings' for the 'living air' made in the workshop. They were embodied in the spontaneous chants for *Sun* from the darkened interiors of the membranes, and in the care invested in every shape and fold. A phenomenology that foregrounds the airy, medial and elemental dimensions of perception aids us in tracing such orientations to air and highlights how bodies are moved, attracted and imaginatively invested in the aerial elements.

Thinking of the perceiving body as always already 'enwinded' or weathered-*in* to an atmospheric milieu is a powerful stance in relation to dichotomies of inside and outside that pervade scholarly and popular discourse.[33] It is a stance that has a long history in feminist materialisms and postcolonial literatures that trouble bodily boundaries in air, water and soil.[34] However, this account of *Becoming Aerosolar* also has purchase for grasping the sensual play and feedback between humans and nonhuman devices and materials. For, it is not only human bodies that are experiencing the force of the sun, the wind and the weather in the event of a launch, and it is not only human subjects who feel elemental spacetimes. The aerosolar membranes, trackers, radar reflectors, enveloped air masses and tethers also respond to or *prehend* their elemental circumstances.[35] A prehension, in Alfred North Whitehead's terms, is a grasping or sensing of one entity by another, or a responding of one entity to another, whether this takes the form of a subatomic particle colliding with a nucleus, a membrane gaining lift with the energy of the Sun, or my looking up at an aerosolar sculpture hovering above me.[36] Whitehead distinguishes between 'positive prehensions', or 'feelings', and 'negative prehensions', 'which are said to "eliminate from feeling"'.[37] Crucially, with this emphasis on prehension, Whitehead explicitly locates *feeling* beyond the human; he suggests human or conscious feelings are 'the exemplification, within our own experience, of a broader kind of process that is far more widely distributed among entities in the world'.[38] Since a prehension is always oriented toward something else, it has a '"vector character"; it involves emotion, purpose, and valuation, and causation'.[39]

More technically, in Whitehead's speculative thought, prehension occurs at levels of reality that are imperceptible to humans. Whitehead uses the term 'actual entity' or 'actual occasion' to refer to 'the final real things of which the world is made up" that are characterised by their prehensions of other actual entities as well as 'societies', 'eternal objects' and 'propositions'.[40] Actual occasions are atomic, ephemeral units of reality that are perpetually emerging and 'perpetually perishing'.[41] As Melanie Sehgal explains, actual occasions are speculative because they are not what we experience; rather, they are *presupposed* by experience.[42] Importantly for my argument here, via processes of prehension, actual occasions stick together, form patterns and coalesce into 'societies' that humans can perceive.[43] Membranes, tethers and radar reflectors are all societies in this sense. Their holding together is an 'achievement' of prehension.[44]

What is happening, then, when the society of occasions that make up the aerosolar membrane is moved and ripped open by the wind? In my reading, the answer has to do with the role of propositions. Propositions introduce novelty. Sehgal writes of propositions, 'when taken up, prehended, they introduce a break into the continuity of a becoming; they divert a historical route and lure it into a different becoming, they generate a different pattern'.[45] Thus, the propositions or lures of the turbulent wind introduce a 'break' in the nexus of actual occasions composing the aerosolar membrane. The entities and occasions of the membrane are lured into a different becoming.

A lure is analogous to a 'propositional prehension' in the sense that it is a feeling, it has a 'vector character' and it introduces novelty; thus, as suggested in Chapter 1, lures are not grasped only by the human body and intellect.[46] James Ash vividly describes the propositional prehensions unfolding between technical objects and atmospheres; however, he calls these *perturbations*.[47] Following Ash's formulation, aerosolar membranes are perturbed (moved, fluttered, levitated) by air, just as the air is perturbed (slowed down, enveloped) by the membranes. In turn, these perturbations produce currents of feeling among entities, matters and media that are registered by human bodies in the atmospheres of the launch. Although it is difficult to reconcile the Whiteheadian inspirations for elemental lures with the object-oriented frameworks that inform Ash's thought, both 'perturbation' and 'prehension' are useful in signalling more-than human responses, lures and feelings.

As Ash points out, accounting for the perturbations of nonhuman entities in elemental atmospheres has implications for control and governance.[48] In other words, how are we to understand the role of political infrastructures governing entities and materials when these entities and materials create ripples and prehensions that are only sometimes registered by humans? Further, how do actual entities, societies and bodies prehend the institutions and laws governing the air? When might a prehension or an entity transgress these laws? In the following sections, I employ another story of a 'failed' aerosolar launch to further develop the elemental lures of wind and weather. More specifically, I attend to the quasi-invisible frameworks governing and regulating the air, frameworks that have featured in more subtle ways earlier in this chapter. Engaging with the social, cultural and political dimensions of air and atmosphere reveals that the atmosphere is a medium for far more than the matter and movement of air. These socio-political conditions and structures also attract, elicit interest and inspire bodies; thus, they are key to the lures of elemental spacetimes.

V The weather of Tempelhofer Feld

As I suggested in Chapter 1, elemental lures are not reducible to naïve appreciations of the beauty and ephemerality of aerial and atmospheric phenomena. Rather, to be lured to the wind and weather, as is the case for participants in the events I have narrated, is to apprehend the social, cultural and political forces suspended in the air. The aerosolar arts stage compelling experiments for tracing these elemental lures. Moreover, they represent a minor or 'nomad practice' because they do not belong to the protocols, initiatives and geopolitical agreements that govern atmospheric space, specifically the troposphere and stratosphere.[49] Unlike corporate modes of aeromobility, the aerosolar arts are not compatible with the economic imperatives of advanced capitalism. They follow the insurgent time of *Aion*, rather than the linear, institutional time of *Chronos*.[50] As such they exist in the gaps, or marginal spaces, of dominant forms of atmospheric politics and governance. For aerosolar practitioners, this situation requires an ability to perceive the frameworks, institutions, laws and policies of the atmosphere, the better to intervene in them. In this way, perception is 'aired' and 'weathered' to include the quasi-invisible yet pervasive presences of aerial governance, law and politics.

This form of perception of the territorialised spaces of the atmosphere is manifested in *A Manual for Becoming Aerosolar* produced by graduate students in the *Becoming Aerosolar* module (co-convened by Tomás Saraceno, Jol Thoms and myself) at IAK in the spring of 2015.[51] The manual was envisioned as a graphic guide to quickly inform new participants in aerosolar practices. To this end, it addressed not only methods for constructing aerosolar sculptures; it also covered the particular forms of perception demanded of this 'nomad' practice. The manual contains chapters on aerosolar history (stretching back to the hypothetical pre-Incan solar balloon and forward to *Museo Aero Solar*); aerosolar form-finding, with particular instructions on how to build an aerosolar tetrahedron and cylinder; a legend for symbols found on plastic bags that index plastic compositions; altitude control and cut-down mechanisms, featuring experiments carried out by Alexander Bouchner, who went on to contribute to *Aerocene*; a chapter on 'aerography'; the volumetric geography of Austrian and German airspace; and a suggestion for employing weather-station infrastructure for aerosolar sculpture tracking, so that such sculptures 'would in fact become part of the ever changing weather conditions'.[52] Already we can detect a meaning of the weather that is more-than-meteorological.

The manual's chapter on Air Law deserves further consideration. It addresses the status of aerosolar sculptures that intend to reach altitudes greater than 30 m and are therefore regulated in Germany under the Air Traffic Act (*LuftVG-Deutschland*) and the Joint Upper Land Aviation Authority of Berlin and Brandenburg (*Gemeinsame obere Landesluftfahrtbehörde von Berlin und Brandenburg*). Important considerations include gaining permission from the owner of the land from which an aerosolar sculpture will be launched; keeping a distance of at least 1.5 km from airports or restricted military areas; and gaining permission from the local aviation authority, especially in circumstances when the sculpture exceeds a minimum weight requirement of 4 kg, when the flight might enter the vicinity of an airport

A MANUAL FOR

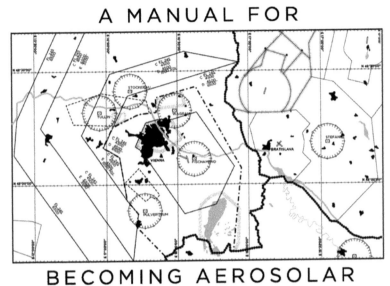

BECOMING AEROSOLAR

Figure 3.4 A Manual for Becoming Aerosolar, 2015; Sasha Engelmann and Jol
Thoms together with Alexander Bouchner, Henry Kirchberger, Lok
Junlin Luo, Jehona Nuhija, Tomi Šoletic, Karla Sršen, Bruna Stipanicic
and Ananda Wie-gandt.

Note: This booklet, published on the occasion of Tomás Saraceno's exhibition *Becoming Aerosolar* at 21er Haus, Vienna, is the outcome of rich and active collaboration carried out at the Institut für Architekturebezogene Kunst, Technical University of Braunschweig, Germany), in the graduate seminar of Tomás Saraceno, Sasha Engelmann and Jol Thoms.

Source: Licenced under CC BY-SA 4.0.

or when the flight occurs in or near a controlled area. In addition, aerosolar sculptures must be visibly marked with the name and address of the 'owner', they must be at least partially reflective in order to be visible by primary radar (although see Chapter 4 for a more detailed account of aerosolar sculptures' 'visibility') and they must be insured. Further restrictions apply on payload composition, ballast and multiple launches. Whether or not one agrees with heavily loaded and exclusionary frameworks like ownership, insurance schemes and the privileging of particular forms of aerial authority, there is no legal way to enter the atmosphere higher than 30 m in Germany without some compatibility to these forms of governance and control.

The navigation of these rules and regulations notwithstanding, the manual reveals an important additional objective of aerosolar practices in relation to law and politics. Immediately after the chapter on Air Law, there is a petition stating:

> Just as engine boats give way to sail boats on international water, so should heavier-than-air vehicles (airplanes, helicopters, etc) give way to solar lighter-than-air vehicles in the air![53]

Drawing on language in the Chicago Convention, the petition suggests a reordering of aerial rights predicated on the value of the least-polluting forms of flying and moving. Since then, this petition travelled to numerous countries under the banners of *Becoming Aerosolar* and *Aerocene* and has been signed by thousands. It conveys an important quality of the politics of aerosolar arts: while intervening in existing codes and regulations of the air, the goal is to change them. For those who come in contact with the aerosolar arts, this gesture proposes that the current forms of governance and regulation of the air are neither tenable nor permanent. It is not likely that the objective of this petition will be met anytime soon. Yet, as a proposition for feeling and imagining, as a lure for the future, the petition is a vital element of aerosolar practices.

If *A Manual for Becoming Aerosolar* is an example of the particular ways that understandings of law, policy and governance are inscribed into aerosolar practices, we must turn to another event of aerosolar launch to understand how these forms of knowledge and perception are played out. Let us consider then, the launch of two aerosolar sculptures from Tempelhofer Feld in Berlin in April 2016. This launch was included in a seminar called Knowing in the Anthropocene convened by Bronislaw Szerszynski, Pablo Suarez, Janot Mendler de Suarez, Melanie Sehgal, Zoe Lüthi and myself, together with Mark Lawrence, Franz Mauelshagen, Tomás Saraceno and Falke Schmidt, as part of the second edition of the Anthropocene Curriculum focused on the Technosphere. The Technosphere is a concept initially proposed by earth scientist Peter K. Haff to describe a distributed sphere of technical capacity including computing power, algorithms, infrastructure and institutions that has grown to be equal in significance to the hydrosphere, biosphere and geosphere.[54] In a critique of this concept, the seminar proposed several questions, namely: Is a different kind of Technosphere possible? What forms of knowledge might a different Technosphere require and engender?

After a day of introductions and discussion, the seminar participants were invited to Tempelhofer Feld (a former military airfield turned park) at dawn. Counter to previous experiences of attempting to launch aerosolar sculptures from Berlin in the month of April, the weather was eerily perfect: there was only a very gentle breeze eddying around the park, and the sun was not filtered by low-hanging clouds. Although the air was cold, the sun quickly warmed exposed surfaces. As sleepy bodies began to assemble on the dewy field, they came with their own forms of investment and attention. Several members of Studio Saraceno including Daniel Schulz were immediately engaged in unpacking the aerosolar sculptures and readying them for launch. Their participation in this event was conditioned by their roles in the economy of Saraceno's studio and their relationship to the studio-produced aerosolar sculptures. In contrast, participants in the Anthropocene Campus seminar approached the sculptures to touch their surfaces but generally remained out of the way of the work that had to be done.

Among the organising team of the Anthropocene Campus seminar, Suarez, Szerszynski and myself had participated in aerosolar events before. At the launch site, our role became one of mediating the relationships between the seminar participants and the action unfurling as the sun rose higher. During the post-inflation period, Szerszynski performed a meditative poem while walking in a circle in the middle of the gathering, as if to encourage the warming membranes to prehend the sun.

Many others had contributed to this launch. Those missing from the event included several who had sewn the fabric sculptures, those who had helped coordinate the HKW and Studio Saraceno institutions, and the IAK students who had contributed to developing the concepts that were activated in this launch. I want to emphasise that aerosolar practices involve uneven forms of investment, expertise and dependence. Following Isabelle Stengers' writing on ecologies of practice, those who participate in such events are 'obligated' to them in different ways: some are more proximate to the physicality of these events; some interface with institutions and bureaucracies; some are more reliant on the economies supporting these launches; and some are co-present in their contributions, yet removed from the event itself.[55] Understanding this is crucial for grasping how a collective of bodies can be co-implicated in elemental conditions and similarly invested in the making of new narratives for elemental being while also situated very differently in social and political worlds.

Tempelhofer Feld was selected as the launch site not only because of its accessibility to the seminar participants staying in Berlin, but also because

Figure 3.5 Two *Aerocene* sculptures warming on the old airport runway at Tempelhofer Feld, Berlin, Germany, 2016.

Note: In the Knowing (in) the Anthropocene Seminar, part of Anthropocene Campus: Technosphere edition convened by Bronislaw Szerszynski, Pablo Suarez, Janot Mendler de Suarez, Melanie Sehgal, Zoe Lüthi and Sasha Engelmann, with Mark Lawrence, Franz Mauelshagen, Tomás Saraceno and Falke Schmidt.

Source: Photography by Justin Westgate, 2016.

the recently abandoned Tempelhofer airfield proved a fitting departure for a consideration of atmospheric politics and more-than-meteorological weather. In his published account of the launch, geographer and seminar participant Justin Westgate notes how the action begins: there was suddenly 'a bustling of activity on this now disused runway'.[56] He continues: 'over the next hour I'm interested to watch the amount of attention these gentle structures require in being nursed to life'.[57] He adds: 'This technology, with its relationship to elemental forces and its sensitivity to conditions, calls for a nurturing disposition'.[58] Observing the disjunctive synthesis between the aerosolar sculptures – 'otherworldly creatures' – and the ongoing presence of military aeronautics in the airport runway – 'the grip of petrochemical power' – he suggests:

> In anthropocenic terms, if we consider the technosphere to be the assemblage of material and immaterial technologies that envelop the surface of the planet, this project inserts a wedge into this to open up other ways of configuring technology and, as a result, our relationships with other entities—both human and other.[59]

Westgate sees the radical proposal of the aerosolar sculptures inflated on the tarmac: they are nomadic entities entering an atmospheric realm dominated by the institutions that created Tempelhofer Feld. He understands that the sculptures are responding to and entering the wind and weather in a mode that is fundamentally different from the heavier-than-air craft for which the airfield was built. Unfortunately for the seminar participants, but interesting for the seminar's considerations of the Technosphere and the Anthropocene, as the two sculptures were about to begin lifting into the air in the beautiful, sunny morning, a van from the local park service arrived. Several people greeted the newcomers. However, the park service employees would not allow the launch to progress. They cited rules governing limits to the size of inflatable objects over Tempelhofer Feld, rules that were unfamiliar to those at the launch. Arguments about the fact that the park normally hosts mass kite launches, small blimps and other aerial objects were made to no avail. The aerosolar sculptures were solemnly deflated. We collectively learned an important lesson: the meteorological weather had been perfect, but the institutional weather was not.

How does one 'see' or 'perceive' the laws, policies and institutions that govern the air, when it is difficult enough to perceive air itself? Like the wind, aerial governance is most palpable when it exerts its full force, but most of the time, especially for those of us living urbanised lives, it remains largely invisible. In this event, however, the laws governing the rights to enter the air over a specific piece of land in Berlin were made visible and sensible. The 'failure' to launch thus catalysed perceptions of other kinds of wind and weather, or, what Andreas Philippopoulos-Mihalopoulos calls the 'expansive *institutional affect*' of the law.[60] The question – *who and what is allowed to enter the air?* – was answered in the abrupt halting of the aero-solar activities.[61] If the park service employees had not shown up, we might have been more oblivious to the particular restrictions in aerial entry rights over the park. The fact that they did come, and the dispute occurred on the 'disused' Tempelhofer runway, served as a metonym for the relationship of the aerosolar arts to the atmospheric Technosphere.

Still, this experience does more than 'make visible' the wind and weather of law and governance. Like many other instances of contestation between aerosolar practices and atmospheric regulations that have arisen over the past years, this experience exposes the air as a heuristic for multiple forms of material, social and political process. Far from a vacuous space or a neu-tral medium, the atmosphere is charged, surveilled, bordered, zoned, mon-itored and regulated. For those intervening in these spaces, perceptions of wind and weather are necessarily interwoven with perceptions of the law and politics of the air. This was epitomised in the moment when the park service employees instrumented their reading of the air law, just as every-one was gathered in a circle, with attentions focused on the two fragile, quasi-amphibious sculptures absorbing the heat of the sun and beginning to lift into the air on an unusually sunny morning. In this moment, my per-ception of the weather at the site suddenly shifted. The faded paint of the air traffic lanes jumped out from the tarmac. I perceived that the former airfield was, legally and politically, still in the air.

VI Aesthesis and poesis of air

The elemental lures of wind and weather emerge, in part, from perceptions of air's elemental materiality and quasi-invisible politics. Thus, an expanded and political phenomenology of air is crucial to elemental lures. The wind coursing over a hilltop in northern Germany is not only the materialisation of pressure differentials and aeolian convection; it is also an imaginative current that has worked its way into an assembly of bodies and a vector that keeps them awake long into the night. It is a political force, too, in the way it gathers, arranges and choreographs bodies and orientations. Similarly, the brilliantly clear, sunny weather of a public park in Berlin permeates bodily states and senses of the future, even when these anticipated futures are frus-trated by what was always already there. Indeed, as evidenced in the history of the aerosolar arts, frustrated launches are as compelling as successful ones in motivating new attempts. Brushing up against, interrogating or pre-hending the borders of governed airspace is often as interesting as drifting through the gaps. These are lures born of aerosolar practices, yet not reduc-ible to these practices. How might a sensitivity to aerial phenomenology or the aesthesis of air lure other practices and experiments into being?

There is a great deal at stake in the perception of air. Indeed, there is an urgency, Eva Horn writes, in 'bringing air (back) to the foreground of our perception as both object and condition of perception'.[62] Alongside its potential value for ethics, a careful investment in the phenomenology of air is useful for tracing the ways bodies are enabled or oppressed by atmospheres of different kinds. In such a project, the quantitative dataset would hold equal value to the social, affective and cultural record of atmosphere. For example, we might locate exposures to air pollutants, such as those that caused the premature death of Ella Kissi-Debrah in the London borough of Lewisham,[63] just as we might trace aero-geographies of racism, like those documented in Simone Browne's *Dark Matters* and Claudia Rankine's *Citizen*.[64] To do so would not be to make an equivalence between toxic particles and toxic affects, but it *would* involve recognising that bodies prehend substances, however invisible, and affective intensities, however fleeting, as accumulative impressions of atmospheres. Furthermore, we might imagine monitoring changes to airspace and air-entry rights together with records of the financialisation of breathing space in cities. We might also imagine documenting local freedoms and barriers to aerial expression. Such projects are already underway through the work of activists, academics and citizens in a range of sites.[65] In different ways, they respond to the elemental lures of contested atmospheres and multiple kinds of weather. The stakes of an aerial phenomenology, and our imaginative capacities to engage elemental conditions, manifest in the intensifying exposures, erosions of rights and ongoing exclusions that unfold, quasi-invisibly, in the air around us.

There are many ways to engage the high stakes of atmospheric exposures and exclusions, however. Together with communities, advocacy groups and critical movements, many scholars in the social sciences and humanities have publicised cases of atmospheric violence and injustice.[66] Without turning away from the obvious importance of this work, we might recognise that a focus on atmospheric violence does not capture the full spectrum of atmospheric experience. Indeed, as Choy and Zee ask, '...in a history of damages, might there lurk other ways of exploring atmosphere?'.[67] One of the contributions of artistic practice is that it can enable ways of apprehending air, atmosphere and weather that are transversal to, yet not innocent of, forms of damage and endangerment. To borrow again from Choy and Zee, this is an apprehension that unfolds through 'the care of building apparatuses of attention elicited by the *lure* of atmospheric sensing'.[68] As apparatuses of attention, aerosolar sculptures and practices manifest the lures of atmospheric sensing and operate differently to documentary scholarship and journalism. A key strength of aerosolar practitioners' engagements with the elemental lures of air and atmosphere lies in an approach that does not diagnose, evaluate or critique as much as it creatively intervenes, provokes and continually questions.

In this chapter, winds and weather systems kept aerosolar sculptures from leaving the trunks of cars, they worked their way into labouring bodies and they stalled rapid ascensions. The wind slashed into the membrane of a levitating aerostat, and the more-than-meteorological weather proved completely unsympathetic for the flight of two others. Although the winds and weather patterns of this chapter have been turbulent, uncooperative and unyielding, they have lured a range of sensual, affective and political experiences and insights. This fact suggests that air and atmosphere have much more to offer the social sciences than is implied in the vertical tracing of borders, the repetitive description of atmospheric immersion or the mere acknowledgement of the spatial circulation of affect. The elemental lures of wind and weather invite further critical studies of the propositions of air and atmosphere. As such, they may contribute to the unearthing and weathering of geographical thought.

Notes

1 T. Ingold, *Being alive: Essays on movement, knowledge and description* (Abingdon: Routledge, 2011), p. 93.
2 T. Saraceno, S. Engelmann and B. Szerszynski, 'Becoming aerosolar: Solar sculptures to cloud cities', in: H. Davis and E. Turpin (eds), *Art in the Anthropocene: Encounters among aesthetics, politics, environments and epistemologies* (London: Open Humanities Press, 2015), pp. 57–62; S. Engelmann, D.P. McCormack and B. Szerszynski, '*Becoming Aerosolar* and the politics of elemental association', in: A. Husslein-Arco and M. Codognato (eds), *Tomás Saraceno – Becoming Aerosolar* (Vienna: 21er Haus, 2015), pp. 67–101.
3 Tomas Saraceno, *Becoming Aerosolar* [exhibition] 21er Haus, Vienna; 21 June 2015–30 August 2015. Available at: www.belvedere.at/en/tomas-saraceno
4 G. Bachelard, *Air and dreams: An essay on the imagination of movement* (E. Farrell and F. Farrell, trans.) (Dallas, TX: Dallas Institute Publications, Dallas Institute of Humanities and Culture, 1988[1943]).
5 T. Choy, *Ecologies of comparison: An ethnography of endangerment in Hong Kong* (Durham, NC: Duke University Press, 2011); J. Zee, *States of the wind: Dust storms and a political meteorology of contemporary China* (Doctoral dissertation, University of California, Berkeley, 2015).
6 For the notion of 'aesthesis of air', see: E. Horn, 'Air as medium', *Grey Room*, 2018, pp. 6–25.
7 B. Anderson and J. Wylie, 'On geography and materiality', *Environment and Planning A*, 41(2), 2009, pp. 318–335.
8 See: D.P. McCormack, *Atmospheric things: On the allure of elemental envelopment* (Durham, NC: Duke University Press, 2018), pp. 152–159.
9 McCormack, *Atmospheric things*, p. 33.
10 J. Edwards and M.B. Long, *Effect of solar radiation upon balloons* (Washington, DC: Government Printing Office: Technological Papers of the Bureau of Standards, 1919).
11 Some stars, planets and black holes are often considered to be near-perfect realisations of black bodies. Interestingly, there is some interest in blackbody-like materials for camouflage, and radar-absorbent materials for radar evasion.
12 It is assumed that air is an ideal gas: the average distance between air molecules is great enough so that the intermolecular forces are negligible.
13 Archimedes' principle is a physical law of buoyancy, discovered by the ancient Greek mathematician and inventor Archimedes, stating that any body completely or partially submerged in a fluid (gas or liquid) at rest is acted upon by an upward, or buoyant, force the magnitude of which is equal to the weight of the fluid displaced by the body.

14 A. Talon and E. Lalique, 'Phaethon, the solar balloon', report written at Vaucan-son High School, France, 2008. Available at: www.vaucanson.org/php5/Accueil/attachments/article/536/phaethon_the%20solar%20balloon_english.pdf
15 M. Serres, *The birth of physics* (Manchester: Clinamen Press, 2000).
16 Serres, *The birth of physics*.
17 Serres, *The birth of physics*, p. 82.
18 Anderson and Wylie, 'On geography and materiality'.
19 T. Saraceno, S. Engelmann and J. Thomson, *Becoming Aerosolar* course descrip-tion (Braunschweig: Institut für Architekturbezogene Kunst, Technical Univer-sity of Braunschweig, 2014).
20 Anderson and Wylie, 'On geography and materiality'.
21 T. Saraceno and A. Krell, communication to students, 2014.
22 Anderson and Wylie, 'On geography and materiality', p. 325.
23 T. Edensor, 'Illuminated atmospheres: Anticipating and reproducing the flow of affective experience in Blackpool', *Environment and Planning D: Society and Space*, 30(6), 2012, pp. 1103–1122.
24 M. Marciniak, personal interview with author, 12 February 2015.
25 E.A. Povinelli, 'Fires, fogs, winds', *Cultural Anthropology*, 32(4), 2017, pp. 504–513.
26 B. Szerszynski, communication to students, 2014.
27 Horn, 'Air as medium'.
28 K. Stewart, 'Atmospheric attunements', *Environment and Planning D: Society and Space*, 29(3), 2011, p. 448.
29 A.F. Whitehead, *Process and reality corrected edition* (New York: The Free Press, 1978[1929].
30 For elaboration of the notion of 'weather-world' and 'weathering world', see: S.A. Schroer, 'On the wing: Exploring human-bird relationships in falconry practice' (Doctoral dissertation, Aberdeen University, 2014); T. Ingold, 'The eye of the storm: Visual perception and the weather', *Visual Studies*, 20(2), 2005, pp. 97–104. See also A. Neimanis and J. Hamilton, 'Weathering', *Feminist Review*, 118(1), 2018, pp. 80–84.
31 Ingold, *Being alive*, p. 88.
32 Horn, 'Air as medium', p. 20.
33 T. Ingold, 'Against soundscape', In A. Carlyle (ed.), *Autumn leaves: Sound and the environment in artistic practice* (London: CRiSAP/Double Entendre, 2007, pp. 10–13.
34 See, for example, L. Irigaray, *The forgetting of air in Martin Heidegger* (Austin, TX: University of Texas Press, 1999); É. Glissant, *Poetics of relation* (Ann Arbor, MI: University of Michigan Press, 1997); M.N. Philip, *Zong! (As told to the au-thor by Sataey Adamu Boateng)* (Middletown, CT: Wesleyan University Press, 2008).
35 Whitehead, *Process and reality*, pp. 19–20.
36 Whitehead, *Process and reality*. Whitehead further explains:

> Prehensions of actual entities – i.e., prehensions whose data involve actual entities – are termed 'physical prehensions'; and prehensions of eternal ob-jects are termed 'conceptual prehensions.' Consciousness is not necessarily involved in the subjective forms of either type of prehension.
>
> (1978[1929]: 23)

37 Whitehead, *Process and reality*, p. 23; For a longer discussion of Whitehead's use and notion of feeling, see: S. Shaviro, 'Whitehead on feelings', *The Pinocchio theory* [blog], np.
38 This is Steven Shaviro's translation of Whitehead's approach to feelings articu-lated in: Shaviro, 'Whitehead on feelings', np.
39 Whitehead, *Process and reality*, p. 19.
40 Whitehead, *Process and reality*.

41 Whitehead, *Process and reality*.
42 M. Sehgal, 'Diffractive propositions: Reading Alfred North Whitehead with Donna Haraway and Karen Barad', *Parallax*, 20(3), 2014, p. 194.
43 Whitehead, *Process and reality*.
44 Whitehead, *Process and reality*.
45 Sehgal, 'Diffractive propositions', p. 196.
46 Whitehead uses the term 'propositional prehensions' in his discussion of the imaginative propositions engaged by historical scholars who speculate on alternative outcomes to the Battle of Waterloo. Although in this example Whitehead is referring to propositions that lure the mental activities of humans, I understand 'propositional prehensions' to apply to the prehension of propositions by actual entities. This is supported in Melanie Sehgals' (2014) reading of propositions through the lens of diffraction. See: Whitehead, *Process and reality*, p. 185.
47 J. Ash, 'Rethinking affective atmospheres: Technology, perturbation and space times of the non-human', *Geoforum*, 49, 2013, pp. 20–28.
48 Ash, 'Rethinking affective atmospheres'.
49 For a discussion of 'minor' or 'nomad' practice and science, see: R. Braidotti, *Posthuman knowledge* (Cambridge: Polity Press, 2019).
50 Braidotti, *Posthuman knowledge*.
51 T. Saraceno, S. Engelmann and J. Thomson (eds), *A manual for Becoming Aerosolar* [publication created in Becoming Aerosolar graduate module by A. Bouchner, H. Kirchberger, J. Luo, J. Nuhija, T. Soletic, K. Srsen, B. Stipanicic and A. Wiegandt] (Braunschweig: Institut für Architekturbezogene Kunst, Technical University of Braunschweig, 2015).
52 Saraceno, Engelmann and Thomson, *Manual for Becoming Aerosolar*, p. 36.
53 Saraceno, Engelmann and Thomson, *Manual for Becoming Aerosolar*, p. 42.
54 P.K. Haff, 'Being human in the Anthropocene', *The Anthropocene Review*, 4(2), 2017, pp. 103–109.
55 Note that the root of obligation, Michel Serres writes, is *ligare*: to bond, to weave. For the 'ecology of practices', see: I. Stengers, 'Introductory notes on an ecology of practices', *Cultural Studies Review*, 11(1), 2013, pp. 183–196.
56 J. Westgate, 'Art, air and ideas in the Anthropocene: Field notes from Berlin', in: Conversations with ACCESS, University of Wollongong, 2016. Available at: www.uowblogs.com/ausccer/2016/05/25/art-air-and-ideas-in-the-anthropocene/
57 Westgate, 'Art, air and ideas', np.
58 Westgate, 'Art, air and ideas', np.
59 Westgate, 'Art, air and ideas', np.
60 A. Philippopoulos-Mihalopoulos, 'Atmospheres of law: Senses, affects, lawscapes', *Emotion, Space and Society*, 7, 2013, p. 36.
61 This question was eloquently posed by Aerocene and Studio Tomás Saraceno member Connie Chester in a talk at the Freize Art and Architecture Summit in London. See: C. Chester, 'What will the future be made of? Tomás Saraceno in Conversation' (with Claudia Melendez, Claudia Chester and Manijeh Verghese) Frieze Art and Architecture Summit London, 4 October 2019.
62 Horn, 'Air as medium', p. 23.
63 S. Laville, 'Ella Kissi-Debrah: New inquest granted into "air pollution" death', *The Guardian*, 2 May 2019. Available at: www.theguardian.com/uk-news/2019/may/02/ella-kissi-debrah-new-inquest-granted-into-air-pollution-death
64 S. Browne, *Dark matters: On the surveillance of blackness* (Durham, NC: Duke University Press, 2015); C. Rankine, *Citizen: An American lyric* (Minneapolis, MN: Graywolf Press, 2014).
65 See for example: N. Calvillo, 'Particular sensibilities', *e-flux Journal*, 2018. Available at: www.e-flux.com/architecture/accumulation/217054/particular-sensibilities/; A. Feigenbaum and A. Kanngieser, 'For a politics of atmospheric governance',

Dialogues in Human Geography, 5(1), 2015, pp. 80–84; Matterlurgy (Helena Hunter and Mark Peter Wright), Air Morphologies, Delfina Foundation, 5 December 2019. Available at: www.delfinafoundation.com/whats-on/air_morphologies/
66 See for example: J. Sze, *Noxious New York: The racial politics of urban health and environmental justice* (Cambridge, MA: MIT Press, 2006); K. Fortun, *Advocacy after Bhopal: Environmentalism, disaster, new global orders* (Chicago, IL: The University of Chicago Press, 2009); A. Feigenbaum, *Tear gas: From the battlefields of World War I to the streets of today* (London: Verso Books, 2017); M. Nieuwenhuis, 'Breathing materiality: Aerial violence at a time of atmospheric politics', *Critical Studies on Terrorism*, 9(3), 2016, pp. 499–521.
67 T. Choy and J. Zee, 'Condition – Suspension', *Cultural Anthropology*, 30(2), 2015, p. 212.
68 Emphasis mine; Choy and Zee, 'Condition – Suspension', p. 216.

References

Anderson, B., and Wylie, J. (2009). On geography and materiality. *Environment and Planning A*, 41(2), 318–335.

Ash, J. (2013). Rethinking affective atmospheres: Technology, perturbation and space times of the non-human. *Geoforum*, 49, 20–28.

Bachelard, G. (1988[1943]). *Air and dreams: An essay on the imagination of movement* (E. Farrell and F. Farrell, trans.). Dallas, TX: Dallas Institute Publications, Dallas Institute of Humanities and Culture.

Braidotti, R. (2019). *Posthuman knowledge*. Cambridge: Polity Press.

Browne, S. (2015). *Dark matters: On the surveillance of blackness*. Durham, NC: Duke University Press.

Calvillo, N. (2018). Particular sensibilities. *e-flux*. Available at: www.e-flux.com/architecture/accumulation/217054/particular-sensibilities/

Chester, C. (2019). What will the future be made of? Tomás Saraceno in Conversation [with Claudia Melendez, Claudia Chester and Manijeh Verghese]. Frieze Art and Architecture Summit London, 4 October.

Choy, T.K. (2011). *Ecologies of comparison: An ethnography of endangerment in Hong Kong*. Durham, NC: Duke University Press.

Choy, T., and Zee, J. (2015). Condition – Suspension. *Cultural Anthropology*, 30(2), 210–223.

Edensor, T. (2012). Illuminated atmospheres: Anticipating and reproducing the flow of affective experience in Blackpool. *Environment and Planning D: Society and Space*, 30(6), 1103–1122.

Edwards, J., and Long, M.B. (1919). *Effect of solar radiation upon balloons*. Technological Papers of the Bureau of Standards. Washington DC: Government Printing Office.

Engelmann, S., McCormack, D.P., and Szerszynski, B. (2015). *Becoming Aerosolar* and the politics of elemental association. In A. Husslein-Arco and M. Codognato (eds), *Tomás Saraceno – Becoming Aerosolar* (pp. 67–101). Vienna: 21er Haus.

Feigenbaum, A., and Kanngieser, A. (2015). For a politics of atmospheric governance. *Dialogues in Human Geography*, 5(1), 80–84.

Feigenbaum, A. (2017). *Tear gas: From the battlefields of World War I to the streets of today*. London: Verso Books.

Fortun, K. (2009). *Advocacy after Bhopal: Environmentalism, disaster, new global orders*. Chicago, IL: The University of Chicago Press.

Glissant, É. (1997). *Poetics of relation*. Ann Arbor, MI: University of Michigan Press.

Haff, P.K. (2017). Being human in the Anthropocene. *The Anthropocene Review*, 4(2), 103–109.

Horn, E. (2018). Air as medium. *Grey Room*, 73, 6–25.

Ingold, T. (2005). The eye of the storm: Visual perception and the weather. *Visual Studies*, 20(2), 97–104.

Ingold, T. (2007). Against soundscape. In A. Carlyle (ed.), *Autumn leaves: Sound and the environment in artistic practice* (pp. 10–13). London: CRiSAP/Double Entendre.

Ingold, T. (2011). *Being alive: Essays on movement, knowledge and description*. Abingdon: Routledge.

Irigaray, L. (1999). *The forgetting of air in Martin Heidegger*. Austin, TX: University of Texas Press.

Laville, S. (2019). Ella Kissi-Debrah: New inquest granted into 'air pollution' death. *The Guardian*, 2 May. Available at: www.theguardian.com/uk-news/2019/may/02/ella-kissi-debrah-new-inquest-granted-into-air-pollution-death

Marciniak, M. (2015). Personal interview with author, 12 February.

Neimanis, A., and Hamilton, J.M. (2018). Weathering. *Feminist Review*, 118(1), 80–84.

Nieuwenhuis, M. (2016). Breathing materiality: Aerial violence at a time of atmospheric politics. *Critical Studies on Terrorism*, 9(3), 499–521.

Povinelli, E.A. (2017). Fires, fogs, winds. *Cultural Anthropology*, 32(4), 504–513.

Philip, M.N. (2008). *Zong!* (as told to the author by Sataey Adamu Boateng). Middletown, CT: Wesleyan University Press.

Philippopoulos-Mihalopoulos, A. (2013). Atmospheres of law: Senses, affects, lawscapes. *Emotion, Space and Society*, 7, 35–44.

Rankine, C. (2014). *Citizen: An American lyric*. Minneapolis, MN: Graywolf Press.

Saraceno, T., Engelmann, S., and Thomson, J. (2014). Becoming aerosolar course description. Braunschweig: Institut für Architekturbezogene Kunst, Technical University of Braunschweig.

Saraceno, T., Engelmann, S., and Thomson, J. (eds) (2015). *A manual for Becoming Aerosolar*. [publication created in Becoming Aerosolar graduate module by A. Bouchner, H. Kirchberger, J. Luo, J. Nuhija, T. Soletic, K. Srsen, B. Stipanicic and A. Wiegandt] Braunschweig: Institut für Architekturbezogene Kunst, Technical University of Braunschweig.

Saraceno, T., Krell, A., Szerszynski, B., Engelmann, S., and Steudte, S. (2014). Personal communications in group discussions during aerosolar workshop at Institut für Architekturbezogene Kunst, Technical University of Braunschweig, 21–22 November.

Saraceno, T., Engelmann, S., and Szerszynski, B. (2015). *Becoming Aerosolar*: Solar sculptures to cloud cities. In H. Davis and E. Turpin (eds), *Art in the Anthropocene: Encounters among aesthetics, politics, environments and epistemologies* (pp. 57–62). London: Open Humanities Press.

Schroer, S.A. (2014). *On the wing: Exploring human-bird relationships in falconry practice*. Doctoral dissertation, Aberdeen University.

Sehgal, M. (2014). Diffractive propositions: Reading Alfred North Whitehead with Donna Haraway and Karen Barad. *Parallax*, 20(3), 188–201.

Serres, M. (2000). *The birth of physics*. Manchester: Clinamen Press.

Shaviro, S. (2015). Whitehead on feelings. *The Pinocchio theory* [blog]. www.shaviro.com/Blog/?p=1309

Stengers, I. (2013). Introductory notes on an ecology of practices. *Cultural Studies Review*, 11(1), 183–196.

Stewart, K. (2011). Atmospheric attunements. *Environment and Planning D: Society and Space*, 29(3), 445–453.

Sze, J. (2006). *Noxious New York: The racial politics of urban health and environmental justice*. Cambridge, MA: MIT Press.

Szerszynski, B. (2014). Personal communication to students at IAK, 21 November.

Talon, A., and Lalique, E. (2008). Phaethon, the solar balloon. Report written at Vaucanson High School, France. Available at: www.vaucanson.org/php5/Accueil/attachments/article/536/phaethon_the%20solar%20balloon_english.pdf

Westgate, J. (2016). Art, air and ideas in the Anthropocene: Field notes from Berlin. In *Conversations with ACCESS*, University of Wollongong. Available at: www.uowblogs.com/ausccer/2016/05/25/art-air-and-ideas-in-the-anthropocene/

Whitehead, A.F. (1978 [1929]). *Process and reality*, corrected edition. New York: The Free Press).

Zee, J.C. (2015). *States of the wind: Dust storms and a political meteorology of contemporary China*. Doctoral dissertation, University of California, Berkeley.

4 Lures of movement

Aerocene Gemini

I Moving with the elements

Cumulus clouds tower in the skies over Lancaster in the North of England on 2 November 2017, as the Mobile Utopia Conference enters its third consecutive day. Throngs of scholars, students and researchers move through conversations about the consequences of mobility infrastructure, drone corridors, geolocation strategies and 'mobile media'. Tethered to a workshop called *The Drift Economy* coordinated by Bronislaw Szerszynski, Grace Pappas and I have organised an *Aerocene* launch. As a crowd of mobilities thinkers and practitioners gathers outside of the conference building, Pappas and I inflate the black *Aerocene* sculpture by running back and forth across a narrow field, holding its mouth open to fill it with air. The sculpture is made of a 15 μm ripstop fabric, and the inflated, bi-pyramidal shape stands out against the glistening grass. Then we tie the payload (consisting of a micro-controller, two sensors, camera and solar battery in a bisected plastic water bottle) onto a corner of the sculpture, securing it with a non-slip knot. As Pappas holds the sculpture, I attach a Garmin Etrex 10 tracking device, a tool that will record the sculpture's movements by plotting a point in space every second. We hold the sculpture as it starts to warm, and after a break in the clouds, the *Aerocene* sculpture begins to ascend.

Since the wind is gusty, reaching nine knots from the northeast, Pappas and I take turns moving around with the sculpture. This requires a high degree of concentration. When holding the tether with our gloved hands, we feel how the membrane reacts to thermal updrafts, lateral gusts, micro-eddies and the lift generated by the warming air inside the envelope. We can sense how the air bends in between the wall of the conference building and the dense row of birch trees at the end of the field. When we notice that the membrane is turning vertically in space, we give it some more line so that it can float higher. When it is pushed sideways like a kite, we hold it as tight as we can. Sometimes we have to lean back against the ropes so that we don't get toppled by the sculpture's momentum. During the next hour, we are almost oblivious to the growing number of people watching, as we move from one end of the field to the other, unable to take our eyes away from the moving membrane. To move in this way is to sense a medium that is itself in motion. It is to dance while holding on to the force of the sun and the air.[1]

In this chapter, I am concerned with elemental lures of transboundary air currents and regulated airspace. I assess and engage these lures by following the movements of airborne entities. A focus on movement implies a willingness to listen to, and be caught up in, the velocities, swerves and vortices of moving matter. Yet as epitomised in the opening vignette, to move with the aerial elements is not to submit to them, but to practice methods of tethering, tracking and tracing. This involves an agile sensing of a material that is animated by forces of planetary and extra-planetary origin. More philosophically, the kind of movement I espouse and seek to detect resonates with Rosi Braidotti's nomadic philosophy and ethics. Although Braidotti does not explicitly engage elemental materialisms and has been noted for her emphasis on biological subjectivity as opposed to more capacious notions of earthen, elemental subjectivity,[2] I follow Braidotti in apprehending movement as 'an inventory of traces'.[3] Braidotti writes:

> The nomad and the cartographer proceed hand in hand because they share a situational need – except that the nomad knows how to read invisible maps, or maps written in the wind, on the sand and stones, in the flora… The desert is a gigantic map of signs for those who know how to read them, for those who can sing their way through the wilderness.[4]

In this passage, Braidotti distinguishes the nomad from the cartographer by the former's capacity to sense the 'invisible maps' of the wind, sand and stone-world, aided by cultural memory and observational skill. Putting aside for a moment the issues with a romantic account of 'reading' the wilderness, Braidotti foregrounds lyric and song as key to nomadic movement. Those who are moved in part by the imagination, she suggests, can move in spaces considered 'blank' or 'barren' by others. The elemental lures explored in this chapter emphasise issues of legibility, politics and the imagination in journeys across the 'deserts' of the upper troposphere and lower stratosphere. Like the ancient forest Marco Polo conjured from a scratch in a chessboard in Italo Calvino's *Invisible Cities*, these movements extend our imaginative sensibilities beyond local context into potentially vast spaces and territories.[5]

For Alfred North Whitehead, thinking is like a flight: 'It starts from the ground of particular observation; it makes a flight in the thin air of imaginative generalization; and it again lands for renewed observation'.[6] This chapter employs a physical, airborne trajectory – a journey in the wind over continental Europe – to think about elemental movement, aeolian currents and contemporary mechanisms of air traffic control. I do so by narrating encounters in the *Aerocene*: an activist and community-based project initiated by Tomás Saraceno in 2015. *Aerocene* has grown into an international network of researchers, artists, scientists and practitioners who employ aerosolar sculptures to query links between fossil fuel extraction, mobility

Figure 4.1 Aerocene launch at the Mobile Utopia conference, 2017, at the University of Lancaster, UK.

Note: Part of the drift economy workshop coordinated by Bronislaw Szerszynski. *Aerocene* launch, coordinated by Sasha Engelmann and Grace Pappas.

Source: Photograph by Sasha Engelmann with aeroglyph drawn by the winds, sun and an *Aerocene* sculpture, photograph and aeroglyph collage by Joaquin Ezcurra.

paradigms and advanced capitalism. In particular, this chapter follows the flight and fate of the *Aerocene Gemini*: a two-part aerosolar sculpture launched from Schönefelde, near Berlin, on 27 August 2016. As they soar to the edge of the stratosphere, the *Aerocene Gemini* will inspire reflections on extended acts of sensing and moving and will summon the stories of other free flights performed between 2016 and 2019. Once they are airborne, the *Aerocene Gemini* will cease to be tangible entities or objects and will become nomadic traces of the elements themselves: *lures of movement* in the air.

II A chase

The *Aerocene Gemini* were launched around 7:30am on a brilliantly sunny Saturday morning in Schönfelde, Germany. There was almost no wind. The sculptures floated a dozen meters high, payloads trailing on ropes below. For some time, nothing happened. The two-part body hung there, like a pair of fragile creatures waiting for change, absorbing the sun's eager rays. 'We are relearning how to float in the air', Tomás Saraceno said.[7] Then, without warning, they caught a draft. A faster current. A line of flight. And they were off, joined together like the ancient twins Castor and Pollux, the *Dioscuri*, half immortal, bodies whose destinies are fatally joined, making their way steadily towards a line of tall trees in the distance.

As *Aerocene* Community member Kotryna Šlapšinskaité related to me, the ascent was gentle. The *Aerocene Gemini* approached the trees, and as the earthbound humans held their breaths, the sculptures barely cleared the tree-tops, payloads intact. Then they disappeared into the blue sky, dissolving into two particles, whose presence had been so palpably felt an hour earlier as two membranous bodies hanging motionless over the field, two delicate creatures sharing a common filamentary web, two aerosols attracting an assembly of prac-titioners, technologies, emotions and gestures in those elemental conditions.

The chase began. A sense of excitement, of thrill, of adventure and equally, of suspense and trepidation. The *Dioscuri* were no longer immediately

Figure 4.2 Michel de Marolles, *Tableaux du Temple des Muses*, Paris, France, 1655, Plate 25.

Note: The plate carries an inscription in Latin from Homer, *Odyssey*, Book 11, lines 303–304: *One day both Dioscuri live, one day they are both dead.*

Source: Photo Warburg Institute. Courtesy of Warburg Institute.

sensible, yet they communicated vital signs: they transmitted location data through the amateur radio-based Automatic Packet Reporting System (APRS), alongside readings of temperature, pressure and humidity, inside and outside their membranes. They captured and forwarded images of the view from their lofty aerial position to a live website where grounded practitioners could follow along. I was one of these grounded practitioners, following the radio-based location of the sculptures, refreshing the SSDV website for new images and speculating on the flight with others. Thus, my sensory realm was extended: it was lured toward pulses of information from those distant, airborne twins, twins whose capacity to pulse such information depended on the 'ecology of practices' that had unfolded on the field that morning.[8] For the chase team and many other distant witnesses, the *Aerocene Gemini* became a tiny red icon of a balloon on the APRS map. Yet the conjoined sculptures were not reducible to the icon, since they continued to engender affective atmospheres, imaginative journeys, geopolitical questions and conversations that played out in breathless proximity in the swelter of a hot summer day.

APRS data forms an 'inventory of traces' employed by the *Aerocene* Community once a sculpture is airborne. In other words, via APRS, the radio spectrum is used to create an 'invisible map' of the air legible to those with the right tools and capacities. APRS is the primary means of sensing a sculpture's movements once it leaves the visible range, enters stronger, higher altitude winds and crosses into airspace employed by heavier-than-air craft. The creation of this map works through a mixture of FM radio signals, transmission protocols, listening stations and digital 'repeaters': an assemblage that, in William Rankin's terms, is 'selectively visible, semi-permanent, and always flirting... with conventional forms of physicality'.[9] A nomadic entity, such as a floating *Aerocene* sculpture, transmits its location and other key information in the APRS mode on a 144.8 MHz carrier wave using an onboard radio transmitter. These transmissions are 'beacons' or 'bursts' that comprise a temporary trace of the sculpture's journey, like breadcrumbs on a trail. As radio amateur Michael Martens KB9VBR explains, there is in fact no expectation that a given APRS transmission is going to be received.[10] Rather, since the beacons are transmitted to every station in range and multiplied by each 'digipeater', the result is that these packets of information are multiplied more than they are lost.[11] An 'IGate' listens to data transmitted through the APRS mode on the radio spectrum 'on air' and 'injects it' into the internet stream, so that live APRS data can be viewed online from anywhere in the world. *Aerocene* sculptures are certainly not the only aerial or mobile objects using the APRS network to geo-locate or to share telemetry and weather information. Rather, viewing the APRS platform map in any location reveals traces of a multiplicity of moving entities, from high altitude balloons to gliders, ships, trucks and passenger cars: a living inventory of wind, water, and fuel-driven nomads.

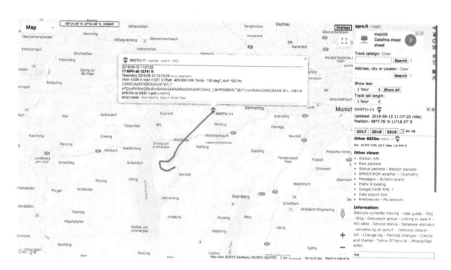

Figure 4.3 Map of the position of *Aerocene 83AQI*, launched on 10 September 2019.
Note: Image captured from APRS platform at 11:07am, when the sculpture reached an altitude of around 3,900 m.
Source: Image capture by Sasha Engelmann, 2019.

Predefined trajectories are virtually impossible in aerostatic flight. The celebrated aeronaut Jeannette Piccard said of balloon journeys: 'You don't file a flight plan; you go where the wind goes. You feel like part of the air. You almost feel like part of eternity, and you just float along'.[12] Yet aero-solar sculptures are even more wayward than most other balloons since they are powered only by the solar energy trapped in their membranes and by the wind that carries them onward. This means that the floating journey of these sculptures is one of 'stillness in motion': moving at the exact speed of the wind, the sculpture joins an elemental current.[13] As such, the traces left by *Aerocene* sculptures on the APRS platform are also 'part of the air'. These can prove strange and surprising to earth-bound practitioners. For example, only 20 minutes after the launch of the *De-NOx* sculpture from the beach of Flakensee on 21 June 2019, the APRS log showed that the sculpture had turned around after reaching several hundred meters in altitude to the west and was coming back over the launch site. Having just celebrated the sculpture's successful launch, those gathered at the lakeshore turned their eyes skyward to see a tiny shimmering particle flying past them in the opposite direction. As it gained even higher altitudes later in the same day, *De-NOx* turned north-eastward and crossed the border into Poland, before returning westward, back into Germany once again. On that day of the summer solstice, the sculpture reached over 21,000 m in altitude: not only was it flying in an altogether non-linear and circuitous route, but it was also surpassing the cruising altitudes at which many fuel burning, heavier-than-air craft fly.[14] In tracing this sculpture's path, then, it becomes possible to observe, even to feel, the curves, patterns and twists of the atmosphere over a particular period of time, as well as the force of the sun in generating the energetic potential for the sculpture's movement. Unlike the elemental lures of shared atmospheres and weather conditions explored in earlier chapters, in aerosolar journeys, the lures of transboundary air movements are communicated in periodic, highly anticipated traces. Each trace-event is a proposition prehended by radios, relays, frequencies, networks, bodies and materials; in these ecologies of prehension, the propositions of the meteorological atmosphere unsettle relations, induce difference and produce novelty.[15]

Although Derek McCormack has argued persuasively that elemental media, from helium to stratospheric winds, are increasingly employed *as infrastructure* by corporate ventures like Google Loon, the experiences of elemental atmospheres I evoke here are marginal to the dominant logistical infrastructures of the air.[16] Yet it is partly this heavily regulated and commercial airspace that lures aerosolar practices. From the early days of aviation, metaphors of road and railroad were employed to describe the pathways of aircraft, as well as the systems of radio navigation used to guide them. This is unsurprising since heavier-than-air flights 'required stable pathways composed of things like radio beacons, rotating searchlights, meteorological facilities, emergency airfields, refueling posts, and wireless telegraphy stations'.[17] Rankin quotes an American airline executive who asserted in 1927 that 'an airway is just as truly on the surface of the earth as is a railway'.[18] Although radio navigation and aviation have advanced significantly since the early twentieth century, practical and legal terminology continues to figure the atmosphere with 'highways', 'lanes' or 'corridors' of movement zoned for heavier-than-air craft.[19] This is the instrumented and logistical air/space in which *Aerocene* intervenes with a very different form of mobility and another relationship to the air. In recent years, members of the *Aerocene* Community have entered into conversation with the European Union Commissioner of Transport, Violeta Bulc. They have proposed that the relatively predictable, south-westerly winds over continental Europe could be employed as 'corridors' of fossil-fuel free mobility.[20] Although the term 'corridor' cannot accurately describe the wind, it is a proposition, a lure, that has the potential to unlock a dialogue with the transportation sectors.

If the atmosphere is morphologised in 'road registers',[21] what are the institutions and practices maintaining these forms of imaging and managing of the air? A substantial body of literature in human geography and critical geopolitics has approached the surveillance of the air through a vertical axis. This work has privileged forms of seeing, witnessing and acting 'from above' in the analysis of military power.[22] However, as Karen Caplan and Weiqiang Lin argue, this literature has missed opportunities for addressing forms of seeing, witnessing and acting from ground to sky.[23] A view 'from below' holds particular challenges: the atmosphere is variably dense and opaque; it is camouflaged and changeable.[24] Yet these challenges to ground-based vision were partly overcome after World War II, when the radar technology that had been employed to sense enemy aircraft was applied to civilian purposes. In radar sensing, aerial entities become visible through transmission and echo: 'An antenna focuses a pulse of energy into a beam of some finite width, and scans the surrounding volume of space... Any reflector within this volume will return some energy back to the antenna'.[25] Radar is fundamental to aerial surveillance today. Still, there are blind spots, shadows and other limits to this form of 'seeing'. After evaluating hundreds of contemporary Air Traffic Management (ATM) manuals, Lin summarises: 'while striving to render the sky as transparent as possible, the *performance* of ATM surveillance is seldom deemed totalising or complete'.[26] Moreover, these forms of technological air-sensing 'overwhelmingly draw on Western – particularly, North Atlantic – logics in aviation systems, surveillance techniques and even weather patterns'.[27] The imposition of North Atlantic 'sky registers' on the rest of the world thus demonstrates a reworking of coloniality under the auspices of air passenger safety. It is a regime that assumes that there is 'one form of sky watching' rather than many.[28]

Although scholars like Lin have engaged with air-sensing and the tracking of heavier-than-air craft 'from below', it is important to consider whether and how moving *Aerocene* sculptures, like the *Aerocene Gemini*, are compatible with conventional forms of aerial surveillance.[29] Contemporary Air Traffic Control (ATC) employs primary and secondary radar systems to sense moving aircraft and other airborne entities. However, ATC's 'primary radar' systems 'consider all non-aircraft targets as undesirable 'clutter,' which must be suppressed to enhance the visibility of aircraft within cluttered or noisy areas'.[30] In many cases, a Moving Target Indicator (MTI) is employed to filter out all non-moving or slow-moving 'radar echoes', such as those from buildings, wind farms, ground vehicles, clouds, insects and trees.[31] While there have been significant improvements since the mid-twentieth century in the detection of 'angels' – or 'unknown and unidentifiable radar echoes'[32] – the resolution of primary radar and the 'targets' of these radar systems mean that a great variety of airborne objects, including clouds, flocks of birds and insect swarms, can pass 'under the radar'.[33] Although *Aerocene* sculptures meet legal requirements for reflective surfaces (they always carry a radar reflector while floating), the dark-coloured envelopes of warm air are not necessarily visible to radar pulses and they move *within* air masses like small clouds. These factors could potentially make *Aerocene* sculptures difficult to distinguish from background 'clutter' in primary radar systems. Moreover, due to their size and weight, *Aerocene* sculptures are not required to carry the transponders obligatory for other aircraft in order to communicate with ATC's 'secondary radar' systems. Without a transponder, even large heavier-than-air craft can go undetected, as reportedly occurred twice in 2014 when Russian military planes turned off their transponders upon entering Swedish airspace, resulting in near collisions with passenger planes.[34] Although it is impossible to determine conclusively whether *Aerocene* sculptures are radar-visible without access to ATC systems, based on available technology and the target-based detection of contemporary aerial surveillance, sculptures like the *Aerocene Gemini* likely hover on the edges of perceptibility for the aviation industry. Like the winds and clouds, *Aerocene* sculptures may barely register on the map.

Let us return then to the *Aerocene Gemini*. The twins reached over 16,000 m in altitude by mid-day. Like *De-NOx*, the *Gemini* crossed the border of Germany to Poland, having caught a strong south-westerly wind. At this point, however, the images loading on the SSDV webpage were either neon rectangles or impossibly granular. I later learned that this was not due to altitude or distance, but due to 'software error'.[35] APRS data was only available via a spot tracker: a manual antenna designed to pick up the signal at closer range. Questions surfaced: where were the sculptures going? What were they seeing? Where would they land? Among the observers, atmospheres of anxiety and worry intensified, caught up in the eerie silence of the sculpture's callsign on the APRS map of continental Europe. Was the flight too ambitious? Would the *Gemini* cross the border to Lithuania, perhaps even Russia? Atmospheres of speculation, anticipation and unrest permeated the group, as the tracing and tracking of the sculptures continued.

Figure 4.4 Aerocene Gemini, free flight, 52°27'32.4"N 14°03'15.3"E–53°54.26'N 21°04.32'E, 2016.

Note: A zero-carbon, *Aerosolar* journey from Schönfelde, Germany to Gmina Biskupiec, Poland on 27 August 2016, with Daniel Schultz, Thomas Krahn, Sven Steudte, Nick Shapiro, Rirkrit Tiravanija, Kotryna Šlapšinskaité, Lars Behrendt, Tomás Saraceno and Alexander Bouchner.

Source: Courtesy *Aerocene* Foundation. *Aerosolar* Photography [onboard camera], 2016. Licensed under CC BY-SA 4.0.

III The lost twins

A hopeful morning was followed by a fearful and uneasy evening. The sun had set, and the sculptures were surely earthbound in Poland, but the signal was lost. It had faded altogether. The chase team, who had driven hundreds of kilometres in the wake of the sculptures, traversed the dark forests and great lakes of Poland near the province of Augustów. It was likely at that point that the sculptures had fallen into a lake or a river, irretrievable. The chasers tried desperately to pick up the signal. A chance reception was the last remaining hope.

The far-away quest for the trace of the sculpture, the search for a signal pulsing weakly through pine, landform, herds of elk, stream, stone and fog, was relayed to those in Berlin via sporadic texts and calls. Bodies shifted nervously, curled into themselves or slumped into grass. There were attempts to distract, refocus, and worry a little less. Yet everyone stared at the little red balloon on the APRS map, marking where it had been sounded last. The evening was punctuated by animated retellings, stories imparting how the sculptures had lifted into the air and the way they had hung, spectre-like, there, until the wind, as in myth, had carried them away.

Braidotti writes of transformations of perceptual coordinates in nomadic methodologies.[36] As the APRS traces were received and decoded, the affective journey of the *Aerocene Gemini* haunted those bound to earth's surface. These hauntings resonated in the electromagnetic alertness of the hand-held radio antenna, poised to receive a signal emanating through the thick darkness of the summer night. Equally, they were found in the packets of temperature, pressure and image-data transmitted from the airborne sculptures over vast expanses of horizon, atmosphere and landscape, picking up textures of these landscapes along the way. The haunting traces were there in the sleepy, strained and sunburnt bodies of those who prepared and launched the sculptures and followed their journey to the edges of the troposphere. Indeed, as Braidotti elaborates, 'The zoe-centred embodied subject is shot through with relational linkages of the symbiotic, contaminating/ viral kind'.[37] As the traces accumulated and then faded, new coordinates and affiliations were established among and between human bodies, landscapes, technologies, weather patterns and movements of air.

I have employed the metaphor of *haunting* because it communicates the spectral quality of this aerosolar journey. By referencing the spectral, I am thinking with scholars like John Wylie, Kristin Gallerneaux and Derek McCormack for whom the spectral is about being sensitive to non-proximate spacetimes and signals.[38] Although I recognise parallels with stories of a 'spiritual ether floating or hovering, wraith-like, above the reassuring solidity of living bodies' the mythopoetic journey of the *Aerocene Gemini* is also spectral because of the traveling of sensation across distant spheres.[39] As the *Gemini* prehended the propositions of photons and the movements of winds at 16 km in altitude, the radio antenna prehended the propositions of APRS data on the 144.8 MHz carrier wave. These propositions on the electromagnetic spectrum were registered by the APRS network of digital repeaters and relays. An unknown number of radio amateurs and mobile entities sensed the far-away trajectory of the *Aerocene Gemini* via its temporary traces on the APRS map. Thus, the elemental lures of transboundary air movements and a quiet passage through regulated airspace were prehended differently by different subjects and entities.[40] These lures of distant aerial movements were embodied, 'admitted into feeling' and became part of the making and sharing of atmospheres on the ground.[41]

As these processes and relations unfolded, the sense of the *Aerocene Gemini* sculptures as discrete objects faded into a sense of their movements. In a way, the sculptures were already 'lost' when they lifted into the air in the morning, since at that moment they became something else entirely: they became part of the currents of wind driven by the displacement of air caused by the uneven heating of the Earth's surface by the sun. During their elemental passage, we could think of the sculptures as alluring entities that were sensible only via technological mediation, thus largely 'withdrawn' from perception and awareness.[42] However, I want to emphasise that the sculptures were simultaneously part of the air, generating and responding to movements outside of human witness or control. The extent to which these sculptures-as-meteorological-movements can be considered objects or entities is therefore limited.[43] Yet, acting at a distance, they continued to produce differences, introduce novelty and move bodies and beings. For my purposes here, after entering the flows of the wind, the *Aerocene Gemini* ceased to be objects or entities and became mobile propositions, or *lures of movement*. If we consider these sculptures as moving lures, we can better grasp how they dissolved into elemental currents, while continuing to carry a host of other bodies and things, from humans to trackers to bursts of data, along with them.

Figure 4.5 *Aerocene* free flight, 52.3923°N, 13.5170°E–53°43'06.1"N 20°05'50.2"E, 2017.

Note: A zero-carbon, *Aerosolar* journey from Schönfelde, Germany to Gmina Łukta, Poland. Together with: Erik Vogler, Sven Steudte, Thomas Krahn, Alexander Bouchner, Desiree Valdés, Sara Ferrer, Sasha Engelmann, Adrian Krell, Kotryna Šlapšinskaité, Sophie Rzepecky, Cara Cotner, Irin Siriwattanagul and Tomás Saraceno.

Source: Courtesy *Aerocene* foundation. Photography by Studio Tomás Saraceno, 2017 licensed under CC BY-SA 4.0.

These lures of movement are further manifested in the relationship between *Aerocene* and local communities of radio amateurs. Since APRS data
is open and free to employ, and since radio amateurs frequently track entities via radio-based methods, radio clubs in Poland have been involved in
tracing and locating *Aerocene* sculptures in response to the ever-shifting currents of wind. One such event occurred during the journey of the sculptures
Cyanophyta I and *Cyanophyta II* launched on 9 June 2018 from Schönefelde.
Cyanophyta I carried a copy of the newly published *Aerocene* book. The two
sculptures were launched at the same time (although without being tethered
together), and rather poetically, like the *Gemini*, they floated side-by-side
for hundreds of kilometres. However, at some point during the journey, the
sculptures separated. *Cyanophyta I* landed in the backyard of an elderly
couple, and *Cyanophyta II* landed further away in a forest.

On this occasion, radio amateurs from Klub Łączności Ratunkowej
SP6ZWR in Poland, namely Przemysław Bienias SQ6ODL, Włodek
Tarnowski SQ6NLN and Michał Lewiński SQ6KXY, participated in locating *Cyanophyta II*. A short film posted by the three radio amateurs on social media documents the minutes before the sculpture was found. The film
shows Bienias and Tarnowski driving through narrow dirt roads in a coni-
fer forest as the sun is setting, talking excitedly while the audible bursts of
APRS, amplified by speakers hooked up to the transceiver on the dashboard,
grow more distinct. One has the sense of a sounding out of something that
is only a burst in the radio spectrum yet is increasingly palpable as something concrete. The trackers drive down an even tinier dirt road, branches
caressing the sides of the vehicle. This leads them meandering among the
pines until they agree they are close enough to walk. As they leave the car,
enter the forest and see the sculpture hanging deflated on a tall pine tree,
they exclaim 'Mamy mamy mamy!' or 'We found it, we found it, we found
it!'.[44] A sense of their elation is tangible through the tones of their voices and
the wobbling of the camera. They give thumbs-up and high-fives to the lens.
After reporting their success to the *Aerocene* Community, they retrieved the
sculpture and its payload (including all reflectors, the GoPro, solar batteries
and tracker) from the tree and brought it to a house where all of the elements
were meticulously weighed and analysed, and damage was assessed before
the entire assemblage was packed and mailed back to Berlin.

Part of the *lure* of *Aerocene* journeys both within and beyond the *Aerocene* Community has to do with the creativity of air's movements. In other
words, despite the existence of aeolian 'corridors', the destination of a particular sculpture cannot be determined in advance. To be sure, a variety
of predictions are employed in preparation for each launch. The *Aerocene*
Community has collaborated with Lodovica Illari, Glenn Flierl and Bill
McKenna in the Earth, Atmospheric and Planetary Sciences Department
at the Massachusetts Institute of Technology to produce the *Aerocene Float
Predictor*: a tool that enables prediction of aerosolar journeys depending
on starting coordinates and wind data derived from live National Oceanic

and Atmospheric Administration datasets.[45] To some extent, this tool of prediction forecasts the lures of transboundary air movements. A variety of other tools, from the Windy App to the Weather Underground site, are employed for wind and weather forecasts. Nevertheless, the path of an *Aerocene* sculpture depends on such a large variety of factors, from the quality of the sculpture's construction and weight to cloud cover, solar radiation and the albedo of Earth's surface, that it is not possible to determine its trajectory with exact precision. In this space of contingency between prediction and outcome, much can occur that has not been anticipated.

In addition to effecting practicalities, such as the chasing of the sculpture or the alerting of local radio amateurs, lures of movement have significant impact on the politics of these practices: in the mutual condition of *not-knowing* that characterises the experience of an *Aerocene* journey, there lies space for imagination, speculation and dialogue. Different forms of data, from altitude to pressure differentials, are openly debated without the possibility of definitive answers. In my experience of *Aerocene* tracking events, while a sculpture is airborne, the overall effect is that of a collective holding of breath, punctuated by periodic revelations when new information is reported. These are experiences of bonding and being in common that might also be called with Timothy Choy and Jerry Zee, 'a work of suspension, of assumptions and disbelief, one that not only describes worlds but holds them in such a way as to allow them to settle into different arrangements, possibilities'.[46] While an *Aerocene* sculpture travels, different positions, affective investments, imaginaries and skillsets enter relationships established and mediated by the aerial elements.

IV Reunion

Circa 11:50pm, a signal emerges. Krell and Krahn hear the APRS burst of the *Aerocene Gemini* on the hand-held radio receiver. They know they are within 3 km. They drive down pitch-dark roads. The signal recedes and they retrace their steps. As they turned down another road, the signal grows louder. Circa 1:18am, the *Gemini* are found. With the antenna in hand, Krell spots the membrane slumped over a bush in a field. The news sparks collective elation. Calls and texts fly among phones at 2am. Some feel intense relief, others pride and others celebrate into the night.

What happened to the *Aerocene Gemini* once they landed? Deflated and flattened, emptied of the warm air and wind that had borne them hundreds of kilometres, the *Aerocene Gemini* transformed. Instead of swerving, hovering and levitating in the air, as they had done that day, the sculptures clung precariously onto twigs and branches. Krell and Krahn had to check for damage that might have occurred as the sculptures came down through the trees. They carried the sculptures from the bushes, then folded and packed them away. Once again, the sculptures became concrete entities with borders, weights, thicknesses and textures.

Figure 4.6 83AQI free flight, 48°02'32.9"N 11°11'03.5"E—49°58'22.6"N 16°17'48.3"E, 2019 22:37 49°58'23.8"N 16°17'49.8"E.

Note: Aerocene recovery team leader Thomas Krahn sensing the *Aerosolar* sculpture 83AQI's presence at the re-landing point in Oucmanice, Czech Republic, 449 km away from the launch site in Munich, Germany.

Source: Courtesy *Aerocene* foundation. Photography by Roxanne Mackie, 2019 licensed under CC BY-SA 4.0.

Although the *Aerocene Gemini* were found deflated, the process of 'grounding' *Aerocene* sculptures is often done manually. In mid-February each year, this activity occurs in a third-year undergraduate module I convene at Royal Holloway University. We spend one two-hour class period preparing, launching (on a tether), floating and ultimately deflating two *Aerocene* sculptures on a rugby field on campus. Although the floating of *Aerocene* sculptures is always thrilling for the students, deflating and packing the sculptures away is often the most surprisingly difficult and memorable. On 13 February 2019, two floating *Aerocene* sculptures were pulled to the ground and deflated by about ten students each. At first unsure how to approach the aerial forms, the students opened the Velcro 'mouth' to let the warm air escape. While some held the corners of the sculpture to the ground, others used their arms to roll the fabric, pushing air out of the opening. Some of them tried to jump on top of the sculpture. Even when the sculpture's membrane was spread flat on the grass, bodies continued to lie, roll and press onto the surface, smoothing out micro-pockets of air. After

Figure 4.7 Deflating an *Aerocene* sculpture, 2019; in 'Atmospheres: Nature, Culture, Politics' undergraduate course convened by Sasha Engelmann at the Department of Geography, Royal Holloway University, London, UK.

Note: *Aerocene* launch coordinated with Jol Thoms and Grace Pappas.

Source: Photograph by Indira Lemouchi, 2019.

ten minutes or so, the membrane was rolled up, folded, and with some difficulty, squeezed back into its bag.

The process of packing and deflating, while seemingly unglamorous, is crucial to the viability of future *Aerocene* launches. Indeed, while it is impossible to precisely predict the trajectories of free-floating *Aerocene* sculptures, the intention is always to recover them so that they may be employed again in the future. To facilitate this sharing and circulating of sculptures, since 2016 the *Aerocene* Community has produced *Aerocene backpacks* as vessels for the membranes and the various tools needed to prepare, launch and experiment with them. These backpacks are about the same size and shape as contemporary urban backpacks and are designed with four flaps that fold in together. The flaps have strips of Velcro and elastic to secure the various elements in place. Although there are minor differences in design and production, each backpack includes an *Aerocene* sculpture packed tightly in a nylon sack; a spool of tether; several pairs of gloves; a manual with instructions on inflating and floating; a payload consisting of Raspberry Pi, solar-cell battery, SD card and USB cable; and a clipboard with an *Aerocene Newspaper Volume 1*.

As a tool for holding *Aerocene* sculptures and technologies together, the backpack is durable and effective. As of September 2019, 39 of these backpacks have been produced and are now shared among users in different cities and countries. In London, for example, two *Aerocene* backpacks are cared for by Grace Pappas and myself, and loaned out to interested individuals. At the same time, the backpack is a design-object that has been exhibited in a number of museums and galleries, including in the V&A Museum in London, the Palais de Tokyo in Paris and the Arsenale Gallery in Venice. As an object on display, the backpack possesses different qualities and lures. It recalls the history of experimental sensing 'kits' created by artists. Like other sensing kits, the display of an *Aerocene* backpack in a gallery comes with particular issues. To suggest what these are, it is helpful to have a specific context in mind.

In September 2019 I travelled to Oslo, Norway, to lead an *Aerocene* 'pilot course' and to help install a backpack in a public exhibition and 'library' on the Architecture of Degrowth for the Oslo Architecture Triennale (26 September–24 November 2019). The library was 'richly populated with drawings, models, materials, artefacts, devices, and ideas exploring the architecture of a Degrowth'.[47] The *Aerocene* Backpack was exhibited in a room with a variety of other experimental and DIY devices. It was attached to a plywood wall with its four flaps open like wings and the various tools displayed inside. A second backpack was also available to be borrowed by the public. Although it is unclear how many people borrowed the backpack in Oslo, the ethic of simultaneous display and lending is crucial for the *Aerocene* Community. However, due to earlier experiences of untrained users experimenting with *Aerocene* sculptures, the capacity to lend a sculpture in Oslo was contingent on a member of the community (in this case, me) being there to offer a 'pilot course'. Because they occupy a place between the airy envelope of the exhibition space and the meteorological atmosphere, *Aerocene* sculptures have an unusual relationship to the gallery context. Yet, lending sculptures requires significant investment from the *Aerocene* Community (it is doubly ironic in the context of the theme of 'degrowth', that I travelled to Oslo by passenger plane). As the *Aerocene* community grows, pilot courses may become more feasible. At the moment, however, the difference between the *Aerocene* Backpack as an entity on display in the tightly regulated airspace of the gallery and the floating *Aerocene* sculpture that dissolves into transboundary air currents raises questions about where *Aerocene* practices are 'consumed', who are they aimed for and what are the ethics of engagement at work in these spaces and formats.

In considering these questions of display and performance, entity and atmosphere, it is vital to attend to issues of knowledge production. In the story of the *Aerocene Gemini*, I described how the mutual condition of 'not-knowing' where the sculptures would land created space for dialogue, the questioning of information and the recalling of experience. These conditions were dependent on the elemental lures of distant currents of air; they were also influenced by how far the sculptures had intervened into the radar-swept territories of European airspace. However, the condition of 'not-knowing', common to *Aerocene* flights, is not only about the impossibility of precisely predicting the sculptures' movements. Since the sculptures

dissolve into elemental air currents, the process of tracking them also requires a degree of humility towards what cannot be fully known. Such humility recognises the limits to our knowledge of the elements, those media
and materials that Astrida Neimanis reminds us are 'much older and far
cleverer' than we are.[48] Hence, in addition to the state of 'not-knowing', the
tracking of a floating *Aerocene* sculpture requires what Neimanis following
Gayatri Spivak calls 'knowing-with' or 'knowing-alongside': being account-
able to what we know, while adopting an attitude of humbled curiosity and
sensitivity towards what we don't.[49]

In contrast, the *Aerocene* sculpture exhibited within the backpack in a
gallery suggests different kinds of knowing-with. In this concrete form, with
the blinking payload tucked alongside the membrane sack, each tool in its
rightful place, ready to be deployed, the backpack references practices of
measurement, observation and imaging. It suggests possibilities of enumerating, navigating and studying air and environment. For some in the community, displaying *Aerocene* sculptures in exhibitions communicates the allure
of the project and its potential usefulness for scientific or citizen-scientific
initiatives. However, for these members of the community, the packaging of
the backpack and sculpture in such displays, its 'entification', can also work
to obscure the 'polyglot working practices' and the many forms of 'invisibal-
ised' labour that sustain the project.[50] In drawing out these distinctions, I do
not wish to moralise about display practices, nor to open up the finer points
of internal discussions in the *Aerocene* community. Rather, I am proposing that the exhibition, display and flight of *Aerocene* sculptures, in which
sculptures are either definable as entities or 'lost' to the elements, map onto
different epistemological, aesthetic and political possibilities. These suggest
different ways of knowing in and about the air and different ways of making
visible the workings of the community. Engaging with issues of knowledge
production through the diverse propositions of *Aerocene* sculptures is vital
to understanding the stakes of the relationship between *Aerocene*, its communities and the art world. Now that we have come back to Earth and more
specifically to institutions of art and architecture, we enter the final twist in
the story of the *Aerocene Gemini*: their return to the atmospheric envelope
of Studio Tomás Saraceno.

V Re-learning how to float

The *Aerocene Gemini* were returned by car to Berlin, where they were unfolded at Studio Saraceno. Videos and data were downloaded. Damage was
assessed. One tiny sensor fell off, probably somewhere over Poland. A few
days later, a gathering was called for all friends of the *Aerocene* to hear a
presentation of the event. In addition to many members of Studio Saraceno,
about a dozen local artists and practitioners attended.

In the series of presentations that ensued, Nick Shapiro, at that time a
summertime *Aerocene* resident and collaborator, talked about open-source

Figure 4.8 Installation view of *Aerocene* in the architecture of degrowth "library"
         at the 2019 Oslo Architecture Triennale, Norway, curated by Matthew
         Dalziel, Phineas Harper, Cecilie Sachs-Olsen and Maria Smith.
Note: An *Aerocene* backpack is displayed unfolded on a wall (left) with *Aerocene Newspaper*,
Vol. 1.
Source: Photograph by Sasha Engelmann, 2019.

licensing and sharing protocols, crucial to *Aerocene* design practices. Sven
Steudte and Thomas Krahn presented the APRS radio transmitters and
camera-boards they had adapted for the *Aerocene Gemini*. Daniel Schulz
spoke about the construction of the sculptures. Then Saraceno made a short
presentation of his own. Instead of highlighting the videos taken from the
sculptures at the edge of the stratosphere, he sped through dozens of images
showing the long drive to the launch site, the unloading of cars, membranes
spread on the field, many smiling faces in various poses, a sunny afternoon at the lake and a scene of waiting in his apartment. He presented the
social-affective texture of the day's experiment: an equal part, he suggested,
of the atmospheric achievement. 'What we are really doing here', he said, 'is
relearning how to float in the air'.

   At the time, the statement struck me as enigmatic. It made me think: when
did 'we' ever float? The directness of the statement, and the all-encompassing
'we', felt too abstract for the constellation of subjectivities and bodies
gathered in the studio that day. I glanced around and saw tired-looking
architects sitting next to production staff with knots of thread still trailing
from their wrists and pockets. I remember thinking that for some people
in the room and beyond it, floating is a euphemism for precarity: more so
than a collective feat, it evokes the inequalities of art work in Berlin at a

time of neoliberal governance, skyrocketing rent prices and the erosion of artist and freelancer collectives. In Elvia Wilk's novel *Oval*, staged in a near future Berlin of artist-consultants, drug-induced generosity and dulled politics, fancy eco-homes quietly float above the city on The Berg, an artificial mountain built on top of Tempelhofer Feld (the old military airfield) that has its own weather.[51] At the American Association of Geographers' conference held in Washington DC in 2019, a three-part session on 'Floating Life' featured presentations on transient living, economic vulnerability and perennial melancholia. The concept of floating, like notions of *weathering*, cannot be dissociated from the politics of location.[52] This was just as true for the post-launch art studio gathering as for other spaces and sites.

In another sense, I later reflected, maybe the statement accesses a longer, deep-time sense of floating. 'We' may have floated long ago, as other forms of life, as bacterial, spore-like, metazoan or amphibian creatures. 'We' may re-learn how to float by experimenting in the principles of buoyancy that would have enabled our ancient ancestors to surf the sulphurous oceans, ride the bubbles of pooling, shadowy rivers or lift temporarily into the air. Indeed, the name *Cyanophyta* is a reference to the earliest micro-organisms that oxygenated Earth's atmosphere.[53] In this sense, then, to re-learn to float would be to engage the ancient, raw and 'inexhaustible' forms of life and creativity that Braidotti calls *zoe*.[54] It would be to re-learn a form of movement more consistent with this *zoe*-centred life, rather than the fuel burning, heavier-than-air and polluting tactics cultivated by *bios*, or contemporary social life. To do so in a way that also respects the politics of location, however, would mean attending to the dynamics of the *Aerocene* Community: to its internal differentiation, its socio- and cultural geographies, its economies of skill and labour. It would mean inventing forms of governance adequate to the project of supporting all of the subjects, matters and entities that join together in this community, an effort that has been the focus of dialogues in the *Aerocene* Forum and in private discussions since the last *Aerocene* Community Summit in June 2019.

At the heart of Saraceno's expression on floating then lies a core issue within the *Aerocene* Community: a question about *Aerocene* as not only a moving practice and a response to the moving lures of the air, but also *a movement*. Indeed, a conversation about the *Aerocene* as 'method' or 'movement' characterised a series of email conversations and posts on the *Aerocene* Forum in 2019, featuring contributions from Camilla Berggren (then a member of the *Aerocene* team in Berlin) and others like Joaquin Ezcurra who work further afield. Embedded in these conversations was the question of whether *Aerocene* is a 'way of doing' atmospheric politics tethered to the specific sensing capacities of *Aerocene* sculptures and kits or whether it is a wider project inventing new imaginaries of post-fossil and 'post-electric' societies.[55] To these questions we might add two parallel ones. Is *Aerocene* a project that privileges some bodies and skillsets over others in the particular geographies of the contemporary art world, further entrenching boundaries and hierarchies? Or is it a movement that respects and encourages difference while accessing a common implicancy in the elements and in the *zoe*-centred histories that archive our undifferentiated origins? In the post-launch event I have described, Saraceno's emphasis on the social textures of the journey of the *Aerocene Gemini* suggests the latter. Yet, as propositions, the answers to these questions cannot be predetermined or solidified: the politics and ethics of *Aerocene as a movement* must be further elaborated through a focus on inner relations of difference and equity, issues that are equal in importance to the sculptures' movements through the 'invisible maps' of the air.

In approaching these questions, the political value of tracking and tracing can be further invoked. Astrida Neimanis employs the figuration of 'the trace' or 'a mark, an impression' as a means to discuss 'the work of politics... that signals the imaginary that one hopes to build and sustain'.[56] For Neimanis, politics is defined in actions, tasks and materialisations needed to confront problems, however difficult this may be. To think of the work of politics as a trace that signals an imaginary is to link the confrontation of problems to imagined alternatives. She continues: 'the 'doing' of an imaginary can't present an imaginary wholesale, but engages the work required to keep negotiating it, and proposing it anew'.[57] Similarly, the 'method' of 'doing' *Aerocene* launches cannot on its own present a post-fossil society. But as a minor confrontation of military and corporate governed airspace, as an unruly proposition of and for the air, the method can offer the means of a continued negotiation of alternative and imagined futures. These 'doings' or political traces, however contingent and malleable, herald an imaginary of fossil fuel-free flight, nomadic wind-born travel, the dramatic reduction of heavier-than-air capitalism and the nurturing of breathable air. At the same time, and more practically, they signal a decentralised, ethics-centred community of aerosolar practitioners and extra-aerosolar networks that is adequate to the task of this future.

For Neimanis following Sara Ahmed, 'to think about politics as a trace that marks or impresses reminds us that these traces work on bodies, situations and worlds'.[58] The 'doing' of any political imaginary, however radical or alluring, leaves traces on those who enact this work. Just as bodies do not float in the same way or to the same degree, so too are they diversely marked by the actions, tasks and materialisations of political projects. We can extend these insights with Rosi Braidotti and the politics of location to suggest that 'an inventory of traces' is not only something floating in the troposphere or propagated on the radio spectrum but also actively moving and pressing on bodies in different ways. The traces of aerosolar journeys are not only registered in sore muscles and in the emotional thrill of the chase; they work on the relations of bodies to subjectivities, bodies to situations (social, cultural, economic) and bodies to more-than-human worlds. Thinking about the inventory of traces in this way means that it is not only the 'invisible map' of a nomadic journey but also the impressions of this journey on those who are trying, in so many different ways, to float. The *lures of movement* of the *Aerocene Gemini* – the way the sculptures entered and became the flows of aerial media, haunting those below – have been elaborated through the arc of this chapter. However, the future of *Aerocene as a movement* depends on a different approach to its method that asks not *whether* a sculpture will fly or *where* it will land, but *how* a nomadic and aerosolar journey can strengthen the bonds between those who are doing the political and imaginative work, thus increasing the buoyancy of the community as a whole.

VI Writing the air

The elemental lures of transboundary air movements and regulated airspaces are prehended in the mythopoetic journey of the *Aerocene Gemini*. Thus, art not only responds to and gets carried away on the restless air, it also intervenes in atmospheric corridors, illustrating possibilities and propositions that lie 'under the radar'. The aerosolar arts conjure meteorological-affective atmospheres of sensing, tracing and discovery, and democratise different forms of aerial knowledge, foregrounding an ethos of knowing-with or knowing-alongside. The shape of this chapter – its narrative arc, shifts in voice and numerous tangents – has been an experiment in *writing-with* these lures of art and the air. While the chapter began on the ground, it moved at different points skyward; it was carried by APRS data and haunted by the spectres of two distant, airborne twins. Indeed, as others have shown, the lures of aerial currents and atmospheric things require a more circuitous, 'slantwise' or 'circumstantial' presentation of material and event.[59] These approaches have more in common with the twisting stories of research and practice than the construction of thesis, demonstration and proof.

Exercises like this attempt to let air's movements 'in' to the page and the argument, bringing other entities and phenomena in too: transponders, insects, reflectors, radar echoes, transmissions and ether as well as hesitation, risk, excitement, anticipation and doubt. This project, which could be called an 'airy poetics' or 'geopoetics', reveals what might otherwise go untraced.[60] It is also an exercise in letting go, allowing the twists and spirals of air to destabilise an unwavering argument. As Luce Irigaray teaches us in her elemental philosophy and poetics, to write with air is not to dismantle or deconstruct but to find the in-between spaces of thought and language, the airy spaces of mutuality and exchange.[61] Irigaray celebrates poets who '[entrust] themselves, immoderately, to that which makes up the body and the flesh of all speech: air, breath, song...'.[62] In doing so Irigaray gives us another elemental lure that has yet to be fully apprehended in elemental geographies or humanistic attentions to air and atmosphere, a lure that challenges academic convention (yet another regulated air/space) and refuses the metaphysics of solids: the proposition of writing air into the page and writing ourselves into the air.

Notes

1 I take the notion of 'holding on to the force of the Sun' from artist, researcher and *Aerocene* Community member Jol Thoms, who first articulated this idea to me and my geography students during a double *Aerocene* sculpture launch at Royal Holloway University in February 2018.

2 See M. Gómez-Barris, *The extractive zone: Social ecologies and decolonial perspectives* (Durham, NC: Duke University Press, 2017), p. 49.

3 R. Braidotti, *Nomadic subjects: Embodiment and sexual difference in contemporary feminist theory* (New York: Columbia University Press, 1994).

4 Braidotti, *Nomadic subjects*, p. 17.

5 Braidotti, *Nomadic subjects*.

6 A.N. Whitehead, *Process and reality*, corrected edition (New York: The Free Press, 1978[1929]), p. 5.

7 I cite this comment here for narrative purposes. This comment was made three days later at an evening presentation at Studio Saraceno at which Saraceno, Daniel Schulz, Sven Steudte, Thomas Krahn, Nick Shapiro and others presented elements of this *Aerocene* journey.

8 I. Stengers, 'Introductory notes on an ecology of practices', *Cultural Studies Review*, 11(1), 2013, pp. 183–196.

9 W. Rankin, 'The geography of radionavigation and the politics of intangible artifacts', *Technology and Culture*, 55(3), 2014, p. 625.

10 M. Martens, 'Introduction to APRS the automated packet reporting system – Ham Radio Q&A', instructional YouTube video. Available at: www.youtube.com/watch?v=xQFSmINZqCY

11 Martens, 'Introduction to APRS'.

12 Jeannette Piccard, cited in K. Gallerneaux, *High static, dead lines: Sonic spectres and the object hereafter* (Cambridge, MA: Strange Attractor Press, 2018), p. 89.

13 To elaborate 'stillness in motion' as a key sensual characteristic of aerostatic flight, Derek McCormack (2009: 33) writes:

> The pioneering aviator Albert Santos-Dumont describes this mode of sensing thus: 'We were off, going at the speed of the air current in which we now lived and moved. Indeed for us, there was no more wind; and this is the first great fact of spherical ballooning. Infinitely gentle is this unfelt movement forward and upward'.

> (Dumont, in Hoffman, 2003: 41)

14 According to Bagshaw and Illig (2019) in *Travel medicine*, a typical passenger aircraft cruising altitude is 30,000 ft or 9,144 m.

15 For a discussion of propositions and novelty, see: M. Sehgal, 'Diffractive propositions: Reading Alfred North Whitehead with Donna Haraway and Karen Barad', *Parallax*, 20(3), 2014, p. 196. However, as Sehgal qualifies: 'The novelty propositions introduce is not necessarily good; no 'theory' is inherently good or bad – everything depends on the situation, the environment and the way a proposition is taken up, *entertained*, and its 'ripples' are prolonged or inhibited' (2014: 200).

16 D.P. McCormack, 'Elemental infrastructures for atmospheric media: On stratospheric variations, value and the commons', *Environment and Planning D: Society and Space*, 35(3), 2017, pp. 418–437.

17 Rankin, 'The geography of radionavigation', p. 628.

18 Rankin, 'The geography of radionavigation', p. 629.

19 See the descriptions of airspace in: W. Lin, 'Sky watching: Vertical surveillance in civil aviation', *Environment and Planning D: Society and Space*, 35(3), 2017, pp. 399–417.

20 My knowledge of these conversations is based on communications with *Aerocene* community members. However, the relationship between *Aerocene* and the EU Commissioner of Transport has been publicised online. See: *Aerocene*, 'Trans-European transport network (European Commission)', *Aerocene*, 2018. Available at: https://aerocene.org/aerocene-to-trans-european-transport-network-european-commission/

21 K. Stewart, 'Road registers', *Cultural Geographies*, 21(4), 2014, pp. 549–563.

22 P. Adey, M. Whitehead and A.J. Williams, 'Introduction: Air-target – Distance, reach and the politics of verticality', *Theory, Culture and Society* 28(7–8), 2011, pp. 173–187; D. Gregory, 'From a view to a kill: Drones and late modern war', *Theory, Culture and Society*, 28(7–8), 2011, pp. 188–215.

23 K. Caplan, *Aerial aftermaths: Wartime from above* (Durham, NC: Duke University Press, 2017); Lin, 'Sky watching'.

24 I. Forsyth, 'Designs on the desert: Camouflage, deception and the militarization of space', *Cultural Geographies*, 21(2), 2014, pp. 247–265.

25 S. Haykin, W. Stehwien, C. Deng, P. Weber and R. Mann, 'Classification of radar clutter in an air traffic control environment', *Proceedings of the IEEE*, 79(6), 1991, p. 746.

26 Emphasis in original; Lin, 'Sky watching', p. 407.

27 Lin, 'Sky watching', p. 409.

28 Lin, 'Sky watching', p. 411.

29 W. Lin, 'Drawing lines in the sky: The emotional labours of airspace production', *Environment and Planning A: Economy and Space*, 48(6), 2016, pp. 1030–1046; Lin, 'Sky watching'.

30 Haykin et al., 'Classification of radar clutter', p. 742.

31 Haykin et al., 'Classification of radar clutter'.

32 Haykin et al., 'Classification of radar clutter', p. 744.

33 J. Yin, C. Unal, M. Schleiss and H. Russchenberg, 'Radar target and moving clutter separation based on the low-rank matrix optimization', *IEEE Transactions on Geoscience and Remote Sensing*, 56(8), 2018, pp. 4765–4780.

34 Associated Press, 'Russian plane has near-miss with passenger aircraft over Sweden', *The Guardian*, 13 December 2014. Available at: www.theguardian.com/world/2014/dec/13/russia-plane-near-miss-passenger-aircraft-sweden; J. Borger, 'Russian ambassador summoned to explain bombers over the channel', *The Guardian*, 29 January 2015. Available at: www.theguardian.com/world/2015/jan/29/russian-bombers-englishchannel-ambassador-summoned

35 T. Krahn, personal communication with author, 2 September 2016.

36 R. Braidotti, *Nomadic theory: The portable Rosi Braidotti* (New York: Columbia University Press, 2011).

37 Braidotti, *Nomadic theory*, p. 224.

38 See for example: J. Wylie, 'The spectral geographies of W.G. Sebald', *Cultural Geographies*, 14(2), 2007, pp. 171–188; D.P. McCormack, 'Remotely sensing affective afterlives: The spectral geographies of material remains', *Annals of the Association of American Geographers*, 100(3), 2010, pp. 640–654; Gallerneaux, *High static, dead lines*.

39 McCormack, 'Remotely sensing', p. 642.

40 Propositions are differently felt by different entities and subjects because they are indeterminate: 'A proposition shares with an eternal object the character of indeterminateness, in that both are definite potentialities *for* actuality with undetermined realization *in* actuality' (Whitehead, 1978[1929]: 258).

41 Whitehead, *Process and reality*, p. 188.

42 For the withdrawn or shadowy qualities of aerostatic entities, consider Derek McCormack's treatment of the giant balloon floating over Manhattan in a story by Donald Barthelme: 'Barthelme's balloon also foregrounds the perplexing

nature of entities as alluring extrusions into worlds whose essence and relations are always beyond us, entities entangled with other entities in ways not always and not necessarily dependent upon human life'. See: D.P. McCormack, *Atmospheric things: On the allure of elemental envelopment* (Durham, NC: Duke University Press, 2018), p. 22.

43 For more on wind as process, force and way of life, see the discussions of 'wind power' in the Isthmus of Tehuantepec: C. Howe and D. Boyer, 'Aeolian politics', *Distinktion: Scandinavian Journal of Social Theory*, 16(1), 2015, pp. 31–48. See also McCormack's discussion of 'atmospheres becoming things' in which he summarises: 'While it may make sense to think of the atmosphere as an entity enveloping the earth, the meteorological processes that characterize this entity are not themselves entities' (2018: 27).

44 Klub Łączności Ratunkowej SP6ZWR, Tak wyglądały poszukiwania dzisiejszego balonu Solarnego ('That was the search for today's solar balloon'). Video, 2018. Available at: www.facebook.com/watch/?v=1719005334815620

45 The *Aerocene* Float Predictor incorporates real-time information from 16-day forecasts of wind speeds at different altitudes. This aerosolar-float trajectory interface is a navigational tool used to plan journeys in the *Aerocene*. Based on a concept by Tomás Saraceno, the *Aerocene* Float Predictor was developed by the *Aerocene* Foundation in collaboration with Lodovica Illari, Glenn Flierl and Bill McKenna from the Department of Earth, Atmospheric and Planetary Sciences at the Massachusetts Institute of Technology (MIT), with further support from Imperial College London, Studio Tomás Saraceno, Radioamateur organisations and the UK High Altitude Society. Atmospheric data is gathered from NOAA's Global Forecast System (GFS), a numerical weather prediction system containing a global computer model and variational analysis run by the US National Weather Service (NWS). The code is open source and available via GitHub: https://github.com/Aerocene/float-predictor

46 T. Choy and J. Zee, 'Condition – Suspension', *Cultural Anthropology*, 30(2), 2015, p. 212.

47 Nasjonalmuseet, 'Oslo architecture triennale: "The library"', 2019. Available at: www.nasjonalmuseet.no/en/exhibitions_and_events/exhibitions/national_museum__architecture/Oslo+Architecture+Triennale%3A+"The+Library". b7C_wRnGlS.ips

48 A. Neimanis, 'Feminist subjectivity, watered', *Feminist Review*, 103(1), 2013, p. 104.

49 See Neimanis, 'Feminist subjectivity'; see also: G.C. Spivak, 'Responsibility', *Boundary 2: An International Journal of Literature and Culture*, 21(3), 1994, pp. 19–64.

50 For a discussion of 'entification' in relation to atmospheric things, see: McCormack, *Atmospheric things*, pp. 24–26. For a critique of display priorities in the *Aerocene* community, see: N. Shapiro, 'Alter-engineered worlds', in: Gary Lee Downey and Teun Zuiderent-Jerak (eds), *Making and doing: Activating STS through knowledge expression and travel* (Cambridge, MA: MIT Press, forthcoming). In this text, Shapiro expands on *Aerocene*:

> Polyglot working practices result from collaborators continuously cross-training each other. That the [*Aerocene*] project is led by an artist likely facilitates interdisciplinarity as no mode of knowing and imagining is prioritized. At the same time, some members and supporters of the project can get frustrated when developing infrastructure, technology, and community take second seat to communicating the potentiality of the idea in art exhibitions.
>
> (Shapiro, forthcoming)

51 E. Wilk, *Oval* (Berkeley, CA: Soft Skull Press, 2019).
52 A. Neimanis and J.M. Hamilton, 'Weathering', *Feminist Review*, 118(1), 2018, pp. 80–84.
53 A.G.J. Þór, 'News story: Free flight at Schönefelde', *Aerocene*, 2018. Available at: https://aerocene.org/free-flight-at-schnfelde/
54 Rosi Braidotti develops the notion of *zoe* and the relationship of *zoe* to *bios* in several of her works, from *Nomadic subjects* (1994) to *Posthuman knowledge* (2019).
55 Shapiro, 'Alter-engineered worlds'.
56 A. Neimanis, *Bodies of water: Posthuman feminist phenomenology* (London: Bloomsbury Publishing, 2017), p. 176.
57 Neimanis, *Bodies of water*, p. 176.
58 Neimanis, *Bodies of water*, p. 176. See also: S. Ahmed, *The cultural politics of emotion* (Abingdon: Routledge, 2013).
59 For example, see: J. Wyatt, S. Tamas and L. Bondi, 'Traces: An introduction to the special issue', *Emotion, Space and Society*, 19, 2016, pp. 37–39; D.P. McCormack, 'Atmospheric things and circumstantial excursions', *Cultural Geographies*, 21(4), 2014, pp. 605–625; K. Stewart, 'Atmospheric attunements', *Environment and Planning D: Society and Space*, 29(3), 2011, pp. 445–453; S. Ballard, L.J. Boscacci, D.S. Carlin, A.A. Collett, E. Hampel, L. Ihlein and T.E. Mitew, *100 atmospheres: Studies in scale and wonder* (London: Open Humanities Press, 2019).
60 For 'airy poetics' see: 'Air's substantiations', in: T. Choy, *Ecologies of comparison: An ethnography of endangerment in Hong Kong* (Durham, NC: Duke University Press, 2011); for 'geopoetics' see: E. Magrane, 'Situating geopoetics', *GeoHumanities*, 1(1), 2015, pp. 86–102.
61 L. Irigaray, *The forgetting of air in Martin Heidegger* (Austin, TX: University of Texas Press, 1999).
62 Irigaray, *The forgetting of air*, p. 156.

References

Adey, P., Whitehead, M., and Williams, A.J. (2011). Introduction: Air-target – Distance, reach and the politics of verticality. *Theory, Culture and Society* 28(7–8): 173–187.

Aerocene Foundation (2018). Trans-European transport network (European Commission). *Aerocene*. Available at: https://aerocene.org/aerocene-to-trans-european-transport-network-european-commission/

Ahmed, S. (2013). *The cultural politics of emotion*. Abingdon: Routledge.

Associated Press (2014). Russian plane has near-miss with passenger aircraft over Sweden. *The Guardian*, 13 December. Available at: www.theguardian.com/world/2014/dec/13/russia-plane-near-miss-passenger-aircraft-sweden

Bagshaw, M., and Illig, P. (2019). The aircraft cabin environment. In J. Keystone, B. Connor, M. Mendelson, P. Kozarsky, H. Nothdurft and K. Leder (eds), *Travel medicine* (pp. 429–436). New York: Elsevier.

Ballard, S., Boscacci, L.J., Carlin, D.S., Collett, A.A., Hampel, E., Ihlein, L.M., and Mitew, T.E. (2019). *100 atmospheres: Studies in scale and wonder*. London: Open Humanities Press.

Borger, J. (2015). Russian ambassador summoned to explain bombers over the Channel. *The Guardian*, 29 January. Available at: www.theguardian.com/world/2015/jan/29/russian-bombers-english-channel-ambassador-summoned

Braidotti, R. (1994). *Nomadic subjects: Embodiment and sexual difference in contemporary feminist theory*. New York: Columbia University Press.

Braidotti, R. (2011). *Nomadic theory: The portable Rosi Braidotti*. New York: Columbia University Press.

Braidotti, R. (2019). *Posthuman knowledge*. Cambridge: Polity Press.

Caplan, K. (2017). *Aerial aftermaths: Wartime from above*. Durham, NC: Duke University Press.

Choy, T.K. (2011). *Ecologies of comparison: An ethnography of endangerment in Hong Kong*. Durham, NC: Duke University Press.

Choy, T., and Zee, J. (2015). Condition – Suspension. *Cultural Anthropology*, 30(2), 210–223.

Forsyth, I. (2014). Designs on the desert: Camouflage, deception and the militarization of space. *Cultural Geographies*, 21(2), 247–265.

Gallerneaux, K. (2018). *High static, dead lines: Sonic spectres and the object hereafter*. Cambridge, MA: Strange Attractor Press.

Gómez-Barris, M. (2017). *The extractive zone: Social ecologies and decolonial perspectives*. Durham, NC: Duke University Press.

Gregory, D. (2011). From a view to a kill: Drones and late modern war. *Theory, Culture and Society*, 28(7–8), 188–215.

Haykin, S., Stehwien, W., Deng, C., Weber, P., and Mann, R. (1991). Classification of radar clutter in an air traffic control environment. *Proceedings of the IEEE*, 79(6), 742–772.

Irigaray, L. (1999). *The forgetting of air in Martin Heidegger*. Austin, TX: University of Texas Press.

Klub Łączności Ratunkowej SP6ZWR (2018). Tak wyglądały poszukiwania dzisiejszego balonu Solarnego ('That was the search for today's solar balloon'). Video. Available at: www.facebook.com/watch/?v=1719005334815620

Krahn, T. (2016). Personal communication with author, 2 September.

Lin, W. (2016). Drawing lines in the sky: The emotional labours of airspace production. *Environment and Planning A: Economy and Space*, 48(6), 1030–1046.

Lin, W. (2017). Sky watching: Vertical surveillance in civil aviation. *Environment and Planning D: Society and Space*, 35(3), 399–417.

Magrane, E. (2015). Situating geopoetics. *GeoHumanities*, 1(1), 86–102.

Martens, M. (2018). Introduction to APRS, the automated packet reporting system – Ham radio Q&A. Instructional YouTube video. Available at: www.youtube.com/watch?v=xQFSmINZqCY

McCormack, D.P. (2009). Aerostatic spacing: On things becoming lighter than air. *Transactions of the Institute of British Geographers*, 34(1), 25–41.

McCormack, D.P. (2010). Remotely sensing affective afterlives: The spectral geographies of material remains. *Annals of the Association of American Geographers*, 100(3), 640–654.

McCormack, D.P. (2014). Atmospheric things and circumstantial excursions. *Cultural Geographies*, 21(4), 605–625.

McCormack, D.P. (2017). Elemental infrastructures for atmospheric media: On stratospheric variations, value and the commons. *Environment and Planning D: Society and Space*, 35(3), 418–437.

McCormack, D.P. (2018). *Atmospheric things: On the allure of elemental envelopment*. Durham, NC: Duke University Press.

Nasjonalmuseet (2019). Oslo architecture triennale: 'The library'. Available at: www.nasjonalmuseet.no/en/exhibitions_and_events/exhibitions/exhibition/national_museum__architecture/Oslo+Architecture+Triennale%3A+'The+Library'.b7C_wRnGIS.ips

Neimanis, A. (2013). Feminist subjectivity, watered. *Feminist Review*, 103(1), 23–41.

Neimanis, A. (2017). *Bodies of water: Posthuman feminist phenomenology*. London: Bloomsbury Publishing.

Neimanis, A., and Hamilton, J.M. (2018). Weathering. *Feminist Review*, 118(1), 80–84.

Þór, A.G.J. (2018). New story: Free flight at Schönefelde. Available at: https://aerocene.org/free-flight-at-schnfelde/

Rankin, W. (2014). The geography of radionavigation and the politics of intangible artifacts. *Technology and Culture*, 55(3), 622–674.

Sehgal, M. (2014). Diffractive propositions: Reading Alfred North Whitehead with Donna Haraway and Karen Barad. *Parallax*, 20(3), 188–201.

Shapiro, N. (forthcoming). Alter-engineered worlds. In Gary Lee Downey and Teun Zuiderent-Jerak (eds), *Making and doing: Activating STS through knowledge expression and travel*. Cambridge, MA: MIT Press.

Spivak, G.C. (1994). Responsibility. *Boundary 2: An International Journal of Literature and Culture*, 21(3), 19–64.

Stengers, I. (2013). Introductory notes on an ecology of practices. *Cultural Studies Review*, 11(1), 183–196.

Stewart, K. (2014). Road registers. *Cultural Geographies*, 21(4), 549–563.

Wilk, E. (2019). *Oval*. Berkeley, CA: Soft Skull Press.

Wyatt, J., Tamas, S., and Bondi, L. (2016). Traces: An introduction to the special issue. *Emotion, Space and Society*, 19, 37–39.

Wylie, J. (2007). The spectral geographies of W.G. Sebald. *Cultural Geographies*, 14(2), 171–188.

Yin, J., Unal, C., Schleiss, M., and Russchenberg, H. (2018). Radar target and moving clutter separation based on the low-rank matrix optimization. *IEEE Transactions on Geoscience and Remote Sensing*, 56(8), 4765–4780.

5 Lures of imagination

D-OAEC Aerocene

I Aerial imaginary

After a fortuitous set of events in October 2018, I found myself on a foggy field at the Aérodrome de la Vallée du Loing, an hour outside of Paris. I was there with two dozen friends and colleagues who were engaged in preparing a series of aerosolar sculptures for launch. In the preceding decade, these colleagues had launched similar sculptures at other sites. They had launched the *D-OAEC Aerocene* sculpture, a 2,973 m³ solar aerostat, designed to *carry a human pilot*, at the White Sands National Monument, in New Mexico.[1] On the dewy field in France that morning, the *D-OAEC Aerocene* sculpture was inflated with air, tended by numerous hands and positioned in the centre of an array of reflective foil strips. Thinking through this event allows me to grasp what lies in air, sun, and cloud, beyond the perception of meteorological movements and towards the uncanny relations of weather and climate.

Our imaginations take flight when we see a face in the clouds, sense a strange presence in a rose-coloured sky or attribute character to the wind. Alexandra Harris demonstrates how metaphor and imagery inform daily experiences of the weather in England.[2] Eliza de Vet uses the term 'weather ways' to describe vernacular practices of adjusting to the weather.[3] In contrast, the climate is often invoked as an abstract or statistical phenomenon that cannot be isolated in singular weather events or individual experiences, no matter how perceptive and imaginative they are.[4] Mike Hulme complicates this notion of climate as 'statistical artefact' when he proposes that climate can be understood culturally as a 'stabiliser' or a 'container' within which weather operates.[5] Moving beyond metaphors of containment or enclosure, this chapter explores other ways of telescoping between weather and climate. I do so by narrating experiments in the aerosolar arts that establish physical and imaginative relationships between bodies, clouds and solar light. In these experiments, bodies become clouds, sculptures become irradiated planets and cloud formations become floating communities. Tethered together, bodies and floating sculptures become something else entirely. The imaginative propositions of art – figurations, images and allegories, or *lures of imagination* – introduce 'a possibility for feeling the world otherwise'.[6] In addition to amplifying perceptions of atmosphere, artistic practices enlarge possibilities for feeling the weather otherwise.

To further elaborate on elemental lures of cloud and sun, especially for discussions of weather and climate, this chapter engages with aerial-elemental constructs and classifications. Tools of elemental classification, from the Beaufort Wind Scale to the International Cloud Atlas, are useful foils for the sensual propositions of art. Yet these classifications are also thoroughly reliant on the imagination. Unpacking their imaginative investments is an exercise in making visible the relations between the aerial imagination, the history of knowledge and institutional power, relations that persist in the present day. If we begin with alternative elemental codes and imaginaries, can we see *beyond* the clouds, the weather and the climate of the colonial and petrochemical era? This chapter considers levitating apparitions, aerosolar heliotropes and atlases of the sky to problematise metaphors of containment and enclosure, investigate inherited material imaginaries and suggest other relationships between bodies and planetary atmospheres.

II Becoming a cloud

As the fog of the early morning lifted, we took turns wiping the dew off the reflective foil, bending rays of light to the under-surface of the *Aerocene* sculpture. *Aerocene* colleagues Jol Thoms and Alice Lamperti then took my hands and guided me into the centre of the launch circle. There, balloon pilot Igor Mikloušić fitted me into a harness and clipped a heavy karabiner

Figure 5.1 Aerocene human flight launches in Paris, France, in 2018; as part of 'ON AIR live with…*Aerocene*', 27 October, in the framework of ON AIR, carte blanche exhibition to Tomás Saraceno, Palais de Tokyo, Paris. Curated by Rebecca Lamarche-Vadel.

Source: Courtesy *Aerocene* foundation. Photography by Studio Tomás Saraceno, 2018 licensed under CC BY-SA 4.0.

onto a loop at my chest. The karabiner was attached to a series of tethers that connected to the mouth of the *D-OAEC Aerocene,* inflated and towering above me. At that moment I could already feel the upward pull, demanding an attention to my footsteps, the tangle of ropes and the sway of the aerostatic body. But the only way to become airborne, at first, was to bounce. Tomás Saraceno helped me tug the tethers to the ground, bringing the sculpture's mouth closer to Earth. Upon release, the sculpture rose once again and the momentum lifted me ever so gently into the humid air.

Throughout this volume I have argued that the aerosolar arts amplify *elemental lures* of shared atmospheres, wind and weather conditions, transboundary air movements and regulated airspaces. In the moments I am narrating here, I experienced another lure, one that was embedded in the physicality of levitating off the surface of the Earth. If a lure is a proposition for feeling, this proposition coalesced in the gradual process of entering the air, pulled upward by the force of the sun. Crucially, this proposition was not contained in my body or mind; rather, it was felt by my body as a range of entities prehended each other, and the sun induced novelty in our circumstances. As Isabelle Stengers argues about the work of Anton Mesmer, when propositions cause ripples in the metaphysical world, they may activate the imagination.[7] While Mesmer's 'magnetic fluid' was impotent in isolation, it became a powerful curative device when animated by the imaginations of those who were 'mesmerised'. For me, the event of floating was not just a vertiginous experience of becoming lighter-than-air but also a shift in the imagined reality of the event itself.

Thinking about lures for feeling helps me reflect in other ways on my experience of the launch. In the moment of levitating off the field, the black sculpture above me and the silver surface underneath, in the midst of beams of sunlight raining down and reflecting up, I experienced a kind of *elemental apparition*. Sky shimmered into land and tree forms. The bodies of friends and colleagues moved around the perimeter of the launch circle in a choreography that felt ritualistic. The strangest sensation was that of being tethered to a massive body of warmed air, held within the black membrane above me. Although it was many times my size, this body did not crush me. It gently levitated, bringing me with it. To be a body composed of saltwater and carbon-based tissue under the Sun is one thing. To be such a body lifted by a *second body* which has absorbed the force of the Sun is quite another. Indeed, the experience of levitating off Earth's surface tethered to a huge membranous body seemed to bend the laws of physics and gravity, even though it was precisely the physics of air and light that enabled the feat to occur.

My experience of lift-off with the *D-OAEC Aerocene* manifests lures for feeling in elemental conditions, and at the same time, a surreal and dreamlike shift in my imagination of these conditions. During the event, the blur and shimmer of my surroundings evoked the optic qualities of a cloud. The feeling of being suspended between earth and sky likely contributed to this cloud-sense. For me, it is not enough to think of this event as a heightened *exposure* to sun, wind and weather. Rather, this event 'clouded' my perceptual anchoring to the conditions in which the launch took place. Although I have since faced incredulity from friends and colleagues, I remain convinced that the experience of levitation enlarged my awareness from the feeling of my body towards feeling of a meteorological entity.

After I unclipped from the harness and returned to the perimeter of the launch circle, others took turns floating, including anthropologist Débora Swistun who featured in Chapter 2. For most people at the launch, the moment that Swistun floated was the most memorable. Indeed, many of the individuals present at the launch had listened to her speech inside *Museo Aero Solar* during the *Aerocene* Symposium at the Palais de Tokyo the previous day. We all felt her words deeply, words that brought into sharp relief the elemental politics of location of *Museo Aero Solar* in the Palais de Tokyo, and the differential ways that bodies apprehend, engage with and are harmed by their elemental circumstances. For participants at the post-launch gathering the next day, Swistun's float became an allegory for another world, one with a different relationship to climatic-affective atmospheres and urgencies.[8] At the time of writing, another significant floating event has occurred in the Salinas Grandes, Argentina, where a school teacher named Leticia Marques, the only registered female aerostatic pilot in the country, piloted the *D-OAEC Aerocene* for several kilometres across the salt desert, and reached over 270 m in altitude on 25 January 2020. This unprecedented, untethered float, known internationally as *Fly with Aerocene Pacha*, was part of a solidarity campaign to protest against Lithium extraction on

Figure 5.2 Aerocene human flight launches in Paris, France, 2018; as part of 'ON AIR, live with...*Aerocene*', 27 October, in the framework of ON AIR, carte blanche exhibition to Tomás Saraceno, Palais de Tokyo, Paris. Curated by Rebecca Lamarche-Vadel.

Source: On Air: Sasha Engelmann. Courtesy *Aerocene* foundation. Photography by Studio Tomás Saraceno, 2018 licensed under CC BY-SA 4.0.

indigenous territories in the region.[9] To grasp the imaginative 'pull' of the worlds engendered by Swistun and Marques' aerosolar levitations, however, it is necessary to further engage with the links between institutional meteorology, aerial knowledge and histories of seeing and sensing the atmosphere. These links are epitomised in the story of the International Cloud Atlas. After tracing the imaginaries, power-systems and practices that informed the Cloud Atlas, we will return to the *D-OAEC Aerocene's* first float in the white dunes of New Mexico to probe the tropes of the aerial imaginary and the interstices of weather and climate.

III Clouds and classifications

A cloud is an aggregate, a nebulous set, a multiplicity whose exact definition escapes us, and whose local movements are beyond observation.

(Michel Serres)[10]

On that October morning at the Aérodrome de la Vallée du Loing, we arrived to a scene of almost complete opacity. A dense fog hung over the field. It was impossible to see more than a dozen metres. Nevertheless the launch team rolled out the reflective foil strips that would bounce any stray photons of light back up into the air and onto the slowly inflating sculpture. Even in these low-light conditions, there were stray energies, electrons to be deflected in their paths. Then, in what felt like a very sudden transformation, the fog lifted to reveal an overcast sky. The *D-OAEC Aerocene* was coaxed into being amidst the clouds, the dew still clinging to its surface. As it took shape, it too became a cloud: it was an 'aggregate' body of warming air particles, a 'nebulous set' held together in a recognisable shape.[11] It was a body highly responsive to its solar and environmental milieu. It is perhaps not surprising then that those who were tethered to the sculpture imagined they were clouds, too.

In order to grasp the allegories, metaphors and figurations of the *D-OAEC Aerocene*, and the complex relationship between the imagination and the aerial elements, it is useful to consider projects that name, define or classify the elements. These projects, for various reasons, create standards, tools and notations to apprehend the elements in systematic ways. Although there are countless examples of attempts to standardise the material-elemental world, from Linneas' species-names to Munsell colour charts to the Periodic Table, my focus here is the classification of clouds as epitomised in the *International Cloud Atlas*, first published in 1896. The story of cloud classification, as scholars like Lorraine Daston have pointed out, is one that stretches the resources of description, and therefore the imagination, to its 'breaking point'.[12] At the same time, it indicates how powerful the lures of elemental constructs can be.

At least since the notations of pre-Babylonian scribes, clouds have been key to the elemental imaginary and to human relationships with elemental phenomena. The Babylonian epic poem *Enuma Elish* describes how the god Marduk transformed various body parts of Tiamat, the goddess of the ocean waters, into the clouds, winds and mists.[13] One of the oldest known representations of a cloud was found at Catal Huyuk, an ancient commercial city on the Anatolian Plateau of Turkey, that shows an amorphous cloud mass hovering over a volcano.[14] In the *Shijing*, the oldest known collection of Chinese poetry, the varying shapes of clouds play an important role in the love story of Zinhü and Niulang, the weaver girl and the cowherd.[15] Chinese porcelain of the Ming dynasty was notable for the advanced process of turning cobalt rock into blue pigment, enabling blue and white inscription. Among many other features, Ming dynasty porcelain is recognised for its detailed cloud forms, often closely associated with lotus flowers, sceptres and dragons.[16]

As the above micro-constellation begins to sketch, clouds have long kindled the analogical, allegorical and figurative imagination. In his book *Cloudland: A Study on the Structure and Characters of Clouds* the nineteenth-century clergyman and meteorologist William Clement Ley documented vernacular interpretions cloud forms by agriculturists and seafaring people of the British Isles.[17] There were as many ways of seeing, reading and imagining clouds as there were towns and villages. Since the seventeenth century, early offices of meteorological observation in Europe had already begun to enlist networks of amateur observers to record clouds and other phenomena in weather diaries.[18] Yet these early systems of meteorological recording, which would evolve into much wider networks in the nineteenth and twentieth centuries, manifested important imaginative investments: in John Ruskin's terms, these networks imagined 'the moving power, of a *vast machine*' that 'wishes its influence and its power to be *omnipresent over the globe* so that it may be able know, at any given instant, the state of the atmosphere on every point on its surface'.[19]

In the late eighteenth century, clouds became objects of European science. Two naturalists Jean-Baptiste Lamarck and Luke Howard offered simultaneous proposals for classifying cloud forms. For several reasons, including Lamarck's use of French rather than Latin and the influence of Napoleon Bonaparte, Howard's classification was more widely accepted. Indeed, Howard's *Essay on the Modification of Clouds* published in 1803 achieved great success in his lifetime.[20] Yet Howard's classification was not revelatory because of its accuracy in categorising cloud-forms, but because of the new *ways of seeing* it presented, based on a classification that mirrored the Linnaean approach to biological life. Howard used names like *cirrus*, *cumulus* and *stratus* to delineate 'certain distinct modifications' of clouds that he likened to 'the countenance of the state of a person's mind or body'.[21] Artists and poets were crucial to the success of Howard's cloud names. Not only did Howard collaborate extensively with the painter Edward Kennion in his renderings of clouds, but Johann von Goethe passionately supported the classification, insisting that cloud observers train themselves not to see 'deviant phenomena' that might confuse the schematic.[22] Goethe's words were prescient: in 1894, less than a century after the initial publication of Howard's manuscript, the newly established International Meteorological Committee led by Hugo Hildebrandsson met in Uppsala, Sweden, to *solidify* and *universalise* the names of clouds. In order to create the first International Cloud Atlas, these meteorologists analysed hundreds of paintings, watercolours and photographs to determine which represented the truth of natural media and which Latin names they should be given. Crucially, to see in so many Latin classifications, these meteorologists had to collectively *un-see* the 'deviations' that lay in between.[23]

While several scholars have investigated the story of the first International Cloud Atlas, a less discussed aspect of this scientific 'achievement' is its complicity with colonial and military infrastructures of the late eighteenth and nineteenth centuries, and the particular imaginaries that informed them. In other words, the 'collective seeing' of cloud forms in Uppsala was also a collective imagining of the sky influenced by colonial and military power. Such imagining of the sky was intimately entangled with the geographical imagination of Empire. We can trace these imaginaries and histories through reports published by the World Meteorological Organisation (the inheritor of the International Meteorological Commission). For example, a 1973 WMO report cites the voyages of Columbus and the associated 'immense stimulus' to European trade as a primary motivator for the standardisation of the science of meteorology.[24] Another passage of the WMO report reads:

> The universal market needed ships in increasing numbers. The security and efficiency of maritime transportation called for precise, reliable and regular information about the weather... It was therefore no accident that the First International Meteorological Conference, which took place in Brussels in 1853, concerned itself largely with maritime meteorological problems.[25]

Indeed, with the exception of a military engineer and a mathematician, all of the twelve delegates of the ten countries who attended the first-ever meteorological conference of 1853 were naval officers. One of the outcomes of this congress was a standard ship logbook with twenty-four columns for notes on wind, the form and direction of clouds, hurricanes, aurora, waterspouts and shooting stars. This logbook, the first product of a successful international standard setting, enabled the systematic observation of the sky.[26] Concurrently, it was a naval and militaristic tool entwined with the particular practices of colonial expansion by sea. Its invention further underscores Ruskin's metaphor of the 'vast machine'; the logbook was a device for amplifying influence *over the globe,* instrumented in part through knowledge of the atmosphere. Naval officers, including members of the US Army and Signal Corps, continued to influence the development of systematised meteorological description, including that of clouds, for many decades to come.

Turning more specifically to the imaginative resources underpinning the classification of clouds, the very possibility of creating an 'atlas' of clouds depended, first, on the assumption that nature was universal and could be idealised to its basic laws and principles by European science, and second, on the reach and extent of observations collected by naval and mercantile forces. In reference to the latter, the meteorologists creating the International Cloud Atlas needed to *imagine* that the same sets of clouds could be observed 'everywhere' on the globe. I use the word *imagine* because without the infrastructures of satellite earth-observation of the twentieth and twenty-first centuries, the imagination necessarily played an essential role in a global theory of cloud forms. Moreover, the meteorologists could not verify their theory without the help of naval expeditions that returned notes, photographs and sketches of clouds. Indeed Reverend William Clement Ley, author of *Cloudland*, interrogated sailors in the British Navy to determine that 'at least four characteristic cloud forms (cirrus, cumulus, stratus, nimbus) were genuinely typical' all over the world.[27] Hugo Hildebrandsson readily acknowledged that the Hildebrandsson-Abercromby cloud classification system adapted from Howard's model and eventually selected for use in the first International Cloud Atlas could not have been made without the expeditions of Ralph Abercromby, who sailed around the world twice in order to ascertain whether the main cloud types could indeed be found 'everywhere'. He was especially noted for his excellent cloud photography, even under the conditions of seafaring. Abercromby, who died in Sydney while undertaking his third around-the-world expedition, came from a long line of military, political and colonial administrators, with links to the British Army, colonial Trinidad, the East India Tea Company and the House of Commons.[28] Thus, the capacity to imagine an 'everywhere' of cloud-forms in order to make an atlas depended on the infrastructures of the imperial navy and tactics of expansion. At the same time, imaginaries of Empire informed the atlas, as much within the minds and bodies of round-the-world explorers as within those of the meteorological committee.

The process of standardisation, Daston writes, is 'an achievement of persons joined in a collective' and 'a prerequisite for a shared world'.[29] However, it is crucial to recognise that the question of who composed and imagined the 'shared world' of cloud and sky classification was historically tied to empire, naval power and coloniality. As Achille Mbembe argues, the European project of expansion since the fifteenth century was an attempt to answer the underlying question: 'who is it that the Earth belongs to?'.[30] The International Cloud Atlas and the infrastructures and imaginaries that fueled its publication and circulation added a second question intimately entangled with the first: *who is it that the sky belongs to?*

New editions of the International Cloud Atlas have since been published irregularly at times when the commissioners feared that cloud descriptions were 'degenerating' into vernacular vocabularies.[31] The present keeper of the International Cloud Atlas – the World Meteorological Organisation (now based in the United Nations in Geneva) – has not added a new type of cloud to the International Cloud Atlas since 1953, despite new proposals by amateur cloud watchers, including Gavin Pretor-Pinney of the Cloud Appreciation Society.[32] Borrowing the words of Stengers once again: 'The approach inaugurated by the commissioners, which has illustrated the critical spirit proper to Science' is one of 'a judgment that manifests the power of reason to dissipate illusion'.[33] In the International Cloud Atlas, clouds, just like other ephemeral fluids and phenomena, are bounded, abstracted and 'brutally reduced' in order to be classified.[34] The International Cloud Atlas is therefore a living record of the judgement and standardisation of elemental media and of the extensive reach of particular imaginaries of an 'everywhere' of cloud description. Even today, the atlas continues to enforce universal imaginaries of clouds. It leaves little room for other forms of sky-knowledge. It also leaves little room for species of elemental phenomena to emerge or disappear, even though our planet's atmospherics clearly manifest such changes.[35]

Without rehearsing arguments of epistemological and cognitive dominance or of the aggression of borders that have been expanded on elsewhere, classificatory schemes like that of the International Cloud Atlas have orchestrated particular relationships between humans and elemental media. They have created 'solid' premises from which to universalise observation and description. These standards are hardly perfect in practice, as Paul N. Edwards shows, but they nevertheless do powerful work as practical, political and imaginative instruments.[36] As Derek McCormack and I wrote, we must better attend to the importance of classificatory diagrams of the elements, 'in shaping a range of associations, some of which are malign and some benign'.[37] A critical approach to elemental classifications must also recognise their legacies and their limitations as rubrics in the midst of a climate emergency. Several artists, including members of the *Aerocene* Community, have proposed counter-atlases of cloud, weather and sky. These counter-atlases seek to recognise multiplicities, deviancies and blurred borders that the International Cloud Atlas ignores. Informing these works are alternative imaginaries of the particular and the universal, the local and the planetary. Before addressing these projects, the following sections turn once again to the imaginative lures of the aerosolar arts with a focus on the first public launch of the *D-OAEC Aerocene*.

IV Aerosolar heliotrope

In my account of the *D-OAEC Aerocene* launch in France, I presented the sculpture, and my experience tethered to it, as a cloud. This is an allegory that has no place in the International Cloud Atlas, despite the fact that Howard's original manuscript made use of bodily metaphors to describe

Figure 5.3 Aerocene human flight launches in White Sands, NM, United States, 2015; performance with the D-OAEC air-fuelled sculpture, successfully carrying passengers without any use of propane for nearly three hours, achieving a World Record for the first, and longest, purely solar-powered, certified, lighter-than-air tethered flight.

Note: The launches in White Sands and the symposium 'Space without Rockets', initiated by Tomás Saraceno, were organised together with curators Rob La Frenais and Kerry Doyle for the exhibition 'Territory of the Imagination' at the Rubin Center for the Visual Arts.

Source: Courtesy *Aerocene* foundation. Photography by Christ Chavez, © 2015.

cloud forms. It is an allegory with a long history that belies another set of imaginative investments in the elements. The envelopment of the body in the clouds permeated accounts of eighteenth-century aeronauts. Several of these early aeronauts were women, including the celebrated Madame Blanchard, who specialised in night-time flights and high stakes launches with firework and flame.[38] Blanchard completed especially long journeys, spending up to 14 hours at a time drifting through the troposphere, pushed by the winds and utterly exposed to the elements.[39] Sadly, the aeronaut took a deathly fall in 1819. A striking engraving published in *Harper's Monthly* in 1869 shows Madame Blanchard's body emerging from a cloud.[40] At the time when Howard's cloud classifications began to circulate in the scientific community, Blanchard was exploring the clouds by physically entering them. One wonders whether, given the platform, Blanchard would have advocated for Howard's system of discrete classifications or might have suggested an entirely different one.

Artists and architects have also probed the relationships between bodies, clouds and climates. In 1958, Buckminster Fuller, working with Shoji Sadao, designed *Cloud Nine*, a series of floating tensegrity spheres, each half a mile in diameter. As Fuller argued in the 1981 book *Critical Path*, a rise in internal temperature of only one degree could levitate an entire cloud community.[41] In 1972 at the height of the OPEC oil crisis artist Graham Stevens staged *Desert Cloud*, an inflatable solar-powered sculpture, in the desert of Kuwait. Will McLean describes the sculpture thus:

> ... the air inside the [*Desert Cloud*] cushion heats up and expands, which then inflates the membrane and the rising heat creates buoyancy, which lifts the whole cloud... the structure could condense or capture water on its surface, miraculously managing to create ice from a clear desert night sky.[42]

Stevens' sculpture was a cloud not only in its physical shape but also in its condensation of water out of thin air. The phrase *Desert Cloud* suggests an anomaly, as well as a form of relief. In contrast, the cloud of *Cloud Nine* was a gathering of humans whose very breath influenced their buoyancy. In these two works, floating entities become allegories for surreal forms of living and levitating. They reveal imaginaries invested in the direct properties of sunlight as it animates matter, the capacity to move without carbon-based fuel, and a techno-fantasy of extreme habitability based not on multiplying technologies but on minimalist climatological application.

These pneumatic 'cloud' sculptures and imaginaries in art and architecture resonate with a series of experiments in solar ballooning that gained momentum in the 1970s. While Stevens' *Desert Cloud* floated in Kuwait, the British architect Dominic Michaelis was constructing a 5,000 m^3 solar balloon that could lift a human passenger. The first solar balloon flight was made in May 1973 by Tracy Barnes in the tetrahedron-shaped *Solar Firefly*. In 1978, the Iranian inventor Frederick Eshoo successfully piloted the *Sunstat*, a half black, half transparent solar balloon, over the white dunes outside Albuquerque, NM. As reporter Dick Brown wrote: '[Eshoo] successfully controlled the *Sunstat's* altitude between 100 and 300 ft by spinning the balloon like a new-born planet rotating in its very own solar system'.[43] This figuration of the balloon as a planet may seem fanciful, but it is less so if we remember that for many decades New Mexico's Tularosa Basin has hosted atmosphere-to-orbital programs including NASA's White Sands Space Harbour and more recently SpaceX and Virgin Galactic. This is a landscape where experiments in flight are yoked to imaginaries of space travel and ideas of the planetary.[44] In 1981, the celebrated balloonist Julian Nott piloted one of Dominic Michaelis' solar balloons (a 3,000 m^3 'double envelope' with a transparent shell and inner membrane) across the English Channel, becoming the first to make the journey in a solar aerostat. When I interviewed Nott at his house in Santa Barbara in 2014, he described how the flight over the channel had been so quiet that he could *hear ships' propellers churning* in the sea below.[45]

The histories of aerostatic flight, solar experiment and climatic architecture choreograph diverse relationships between humans and elemental air. It is therefore significant that on 8 November 2015, only a few minutes' drive away from Eshoo's launch site, a team led by Tomás Saraceno launched the *D-OAEC Aerocene*. The launch was supported by a symposium at the Rubin Centre for the Visual Arts in El Paso called *Space Without Rockets* (curated by Rob La Frenais and Kerry Doyle) and formed part of the project *Territory of the Imagination: At the Border of Art and Space*. John Powell, founder of JP Aerospace and author of *Floating to Space* (a book about inflatable high-altitude platforms), was present at the launch. Daniel Schulz and Danja Burchard of Studio Saraceno were central in the performance. Curator and researcher Nicola Triscott was among those who floated. She evokes the relationship between the *D-OAEC Aerocene* and the vertiginous spaces of the atmosphere as she recalls her experience drifting over the white sands: '[I] felt the immediacy of the possibility of flying up and off in whatever direction – and across whatever border – the elements might choose, should my colleagues on the ground let go of those two slight ropes'.[46] After returning to Berlin, Saraceno recounted how the upward *pull* of the *D-OAEC Aerocene* over the white sands had been so strong that it felt like skydiving but *away from Earth's surface*.

   This spectacular event, captured in powerful imagery that circulated wide and far (see Figure 5.3) and featured in the *Aerocene Newspaper, Volume 1* published for an exhibition at the Grand Palais during the COP21 Climate Conference in Paris, has been described as a feat of human ingenuity and 'unbounded creative spirit',[47] new 'thermodynamic models'[48] and resistance to the privatisation of space flight.[49] However, these readings largely miss the historical continuity of solar aerostatic flights that links the *D-OAEC Aerocene* launch to previous experiments in White Sands and other similar landscapes. Indeed, the *D-OAEC Aerocene* is connected to Dominic Michaelis' early solar balloons in more than its shape.[50] Yet, this continuity is not only one of invention; there are important threads in imaginaries of acclimatisation (*Desert Cloud*; *Cloud Nine*) and the abundant 'gift' of solar light (*Sunstat*; *Solar Firefly*). The imaginative propositions of the *D-OEAC Aerocene* are best understood as part of this historical lineage rather than as a break or departure from them.

   In New Mexico the upward pull of the *D-OAEC Aerocene* sculpture was so strong in part because of the high amount of reflective radiation from the white sand. Paying closer attention to the sand opens up another, equally important historical and imaginative context. As Vanessa Agard-Jones writes in *What the Sands Remember*, the black, grey and mottled sandy beaches of Martinique offer testimony to the 1902 eruption of Mount Pelée that killed tens of thousands of people living in the city of Saint Pierre.[51] Christina Sharpe cites Agard-Jones when she reflects on the sands of the Sahara Desert (another *sea* migrants must cross on their way to the Mediterranean) and the soil memory of sites where lynching occurred in America.[52] The sands of New Mexico remember other stories. These are stories vividly narrated by Japanese writer Kyoko Hayashi, whose novel *From Trinity to Trinity* offers a semi-autobiographical account of the journey of a Nagasaki bomb survivor to the Trinity Test Site, just north of the White Sands National Monument, where the first atomic bomb was detonated in 1945, wrenching apart the earth, plant and animal life and indigenous settlements of the Jornada del Muerto with the fission of Plutonium-239.[53] Even in old age, Hayashi's protagonist is lured to the desert landscape because for her, the bomb has not stopped exploding: '[the] cascading energies of the nuclei that were split in [the] atomic bomb explosion live on in the interior and exterior of her body'.[54] In her attempt to *find new coordinates* of mourning and remembrance, Hayashi's protagonist encounters pueblos, plants, animals, earth elements and the radioactive material *Trinitite*, also survivors (*hibakushas*) of atomic violence.

Nuclear culture scholars have long pointed out that American Cold War propaganda likened the Trinity Test and other nuclear detonations to feats of harnessing the power of the sun via physicochemical nuclear processes. This solar allegory lurks in proposals made by geologists like Colin Waters and Jan Zalasiewicz that the Plutonium-239 fallout from this and other atomic events is a possible 'golden spike' or 'marker' for the Anthropocene epoch.[55] A thorough engagement with the imaginative lures of the Anthropocene 'spike' for Euro-American intellectual cultures is beyond the scope of this chapter.[56] Nevertheless, it is important to recognise that the flights of the *Sunstat* and the *D-OAEC Aerocene* among others occurred in a material, historical and imaginative matrix of the Atomic Age, solar light, corporate space travel, indigenous genocide, the Anthropocene, the sand and the US-Mexico border. They also occurred in spatio-temporal proximity to the most violent cloud in history: the *mushroom*.

Turning more specifically to the *D-OAEC Aerocene*, we should ask: what are the imaginative lures of this launch at the White Sands National Monument? Is it the vertiginous image of an atmosphere-to-orbital technology that does not emit black carbon as suggested by 'space without rockets'? Is it another extraterritorial allegory of floating to other planets? I prefer to think with Elizabeth DeLoughrey in proposing that the *D-OAEC Aerocene* participated in imaging and imagining the intersections of nuclear ecology, solar radiation and the instrumentation of light that we might call a *heliotrope*.[57] For DeLoughrey the heliotrope invokes the alterity of light as that which illuminates but eludes rational grasp. Relatedly, in our work on elemental aesthetics, Derek McCormack and I wrote: 'the sun is the elemental entity on whose alluring emanation we depend most but the essence of which remains absolutely beyond us'.[58] At the White Sands National Monument, the uncanny dimensions of light meet the elementality of the sun in a landscape that has felt the full force of physicochemical nuclear processes.

In photographs of the *D-OAEC Aerocene* launch, the black membrane of the sculpture stands out against the shimmering white landscape. In this heliography, the dark membrane is an anomalous cloud, an absorbent cavern of trapped light, as well as an 'energy-converter' transmuting light to heat.[59] Where is the human body? For DeLoughrey as for Hayashi, the body is a radiant entity too. After thousands of nuclear tests worldwide, especially the series of thermonuclear explosions in the air and seascapes of the Marshall Islands, radioactive isotopes of carbon and strontium have been absorbed by the teeth and bones of all humans on the planet. Therefore, the body of the pilot levitating over the white dunes is also a heliotrope, a figure of light in the nuclear age. If we follow the elemental lures of clouds and solar light and access a heliotropic imaginary, the *D-OEAC Aerocene* launch cannot be reduced to or celebrated as utopian fabulation or space-era innovation. Like the sand itself, the imaginative lures of this event are dis-figuring, inside and out; they '[get] inside our bodies, our things, in ways at once inconvenient and intrusive'.[60] In this process of dis-figurement our bodies become strange vessels, tied to a planet illuminated and irradiated by so many suns.

Thinking of the launch of the *D-OAEC Aerocene* as a heliotrope enables a different kind of 'responsible literacy' of the White Sands launch and its spectacular imaging.[61] In other words, this attempt to read the imaginative lures of the launch through cloud and light complicates interpretations of the event that offer immediate comprehensibility and are too quick to describe an act of heroism. Moreover, thinking with the figures and imaginaries of the heliotrope troubles the position of the reader or analyst; we can no longer view ourselves as objective interpreters, distanced observers or agents. Instead, as suggested in the dis-figuring of the body as a radiant entity, we are creatures defined by our relations to the inexhaustible difference and alterity of the planet. This, DeLoughrey writes, is what Spivak calls 'planetarity':

> Planetarity, [Spivak] argues, insists that we configure our relation to alterity as not necessarily continuous nor discontinuous. It is the process by which the familiar is rendered uncanny and unhomely, similar to the ways that the apprehension of (invisible) radiation and its ecological properties destabilize our understanding of place and space. Planetarity is a method of reading that represents 'the de- familiarization of familiar space' just as our recognition of the physics of light foregrounds its 'uncanniness' as it 'puts us in touch with distant, seemingly untouchable entities'.[62]

Planetarity is thus a method of reading and interpreting that coalesces in the apprehension of the uncanny physics of light. In reading the launch of the *D-OAEC Aerocene* as a heliotrope, we are translating the familiar image of a balloon launch in unfamiliar terms. By focusing on its strangeness, we apprehend our own. The image of the single pilot levitating under the black membrane in front of the glittering white dunes, trailed by tiny dots of human bodies below, radiates into our bodies and our imaginaries of place and planet, collapsing and twisting the optics of interpretation. Hence, the event telescopes between the dark envelope and 'nebulous set', photonic ray and the planetarity of light, the mind-body of the reader and the planetary

Figure 5.4 Aerocene human flight launches in Paris, France, 2018 as part of 'ON AIR, live with...*Aerocene*', 27 October, in the framework of ON AIR, carte blanche exhibition to Tomás Saraceno, Palais de Tokyo, Paris. Curated by Rebecca Lamarche-Vadel.
Source: Photography by Mark Neyrinck.

'other'. These propositions for 'becoming planetary' can further inform a reading of weather and climate via the levitating body and the uncanny experience of floating with the clouds and the sun.[63.]

V The second body

The images taken in the last hour of the *D-OAEC Aerocene* launch in France, once the fog had lifted, show a constellation of buoyant bodies. The silhouettes of the *D-OAEC Aerocene* and its human pilot are clearly visible, but so too are three smaller *Aerocene* sculptures. One of the most beautiful aspects of those moments was the relational movement of the sculptures with a human body suspended in their midst. As winds blew, clouds cast shadows and the sunlight shifted direction, this nebulous constellation moved, too. There was the sense of a different language, one transmitted in cascades of photons, the formations of clouds and vectors of air. There was also a sense of being proximate with 'seemingly untouchable entities'.[64] In the arrangement of bodies, it was as if the clouds and the Sun, like the winds and the membranes, were moving as parts and as a communicating whole.

I have proposed that the physicality of levitating with the *D-OAEC Aerocene* conjures experiences of elemental bodies and entities, like those of clouds. This is another way of asking the question: where does the body stop and the weather begin?[65] I have also argued that *Aerocene* launches are not only technical feats, but also imaginative allegories of cloud and solar light with relationships to particular landscapes and histories. Thinking with these allegories can thus inform another question: where does the weather stop and the climate begin? From the insights generated in this chapter so far, we can begin to grasp that the imaginative lures of the aerosolar arts may trouble separations or 'containers' for weather and climate. To attend to these questions further, we need to return to the body.

The Shakespeare scholar Daisy Hildyard suggests that each human has *two bodies*. Your first body is the body, 'you live in, made out of your own personal skin'.[66] This is the body that sleeps and wakes, goes to the corner shop for milk or eats breakfast standing up in the kitchen. Your *second body* 'is not so solid as the other one, but much larger'.[67] This *second body* is your own existence as a force on the climate, a change in bio-, atmo- and geo-spheres and any number of other processes that are part of these changes. Hildyard writes:

> Your first body could be sitting alone in a church in the centre of Marseille, but your second body is floating above a pharmaceutical plant on the outskirts of the city, it is inside a freight container on the docks, and it is also thousands of miles away, on a flood plain in Bangladesh, in another [wo]man's lungs.[68]

The trope of the body is employed to construct an allegory of a *second* body, a physical-elemental expression of matter, energy and influence. The power of this allegory lies in its doubling of the familiar (the body) with the un-canny other (the second body). Like a nebulous cloud or solar light, the second body possesses an alterity or 'planetarity' that is both continuous and discontinuous with the (first) body.[69] This is an allegory and a set of material relationships that ripple between scales, forcing us to consider how these bodies and scales are defined in the first place.

To further unfold the notion of the second body as an allegory for plan-etarity, it is helpful to think in cloud terms. The body of a cloud has a certain weight and form. It moves around. It associates with other cloud bodies. At the same time, the body of each cloud permeates the body of every other cloud, both literally, as water droplets circulate in the hydro-logical cycle, and more abstractly, as each cloud-form generates energetic relationships that alter the climate system for all other clouds. Moreover, today we are used to thinking about the virtual 'cloud' of cloud computing, whose patterns of distribution in the humming of data centres, server farms, fibre-optic infrastructure, content storage platforms and high frequency re-lays are amplifying warming atmospheres and rising sea levels. To think of clouds in this way, as having at least two bodies, is to depart from the classificatory codes of the International Cloud Atlas, in which clouds only ever have one. Following the words of Serres cited earlier, this is an experi-ment in the becoming-multiple of clouds. It is also a proposition for holding together weather and climate in ways that do not overlook the value of these terms but also refuse to let their normative meanings dictate possibilities of imagination and critique. Whitehead remarks: 'it is more important that a proposition be interesting than that it be true'.[70] In other words, it is more important that a proposition have 'relevance' for thinking, doing and feeling than that it conform to pre-established truth claims or frameworks.[71] For example, a proposition for holding together weather and climate might go as follows: if a cloud can have a second physical body with trans-scalar and temporally extensive qualities, then the cloud that is contributing to the weather conditions of an aerodrome in France is not limited to a classificatory container from which it can't escape. It is simultaneously the climate, held together in the shape of a cloud.

The allegorical and imaginative leap to the second body is greater, perhaps, for humans than for clouds. Nevertheless, the second body has already figured prominently in this chapter, in my perceptions of becoming-cloud during the *D-OAEC Aerocene* launch. Indeed, Hildyard insists on the physicality of the body of the individual as a way of accessing the second body. It might be, therefore, that the key to the imaginative lures of clouds and solar light is not only the act of levitating up, but also the act of 'coming down' to skin, flesh and bone. It is by turning to the body that we can confront totalising rubrics as well as models that render Earth a transparent globe under an Apollonian eye. It is also through the body that we might find other ways of relating weather and climate that do not begin from statistical limits or singularities, but instead conceive of the climate as an uncanny figuration of otherness and elsewhere, held metonymically within every change of weather. In Whitehead's terms, what 'relevant' and 'interesting' possibilities would such perspectives offer at a time when the nexus of weather and climate pose such powerful consequences for human existence and politics?

Figure 5.5 Tomás Saraceno, *Aero(s)cene: When breath becomes air, when atmospheres become the movement for a post fossil fuel era against carbon-capitalist clouds*, 2019; installation view of *On the Disappearance of Clouds*, 2019 at the 58th International Art Exhibition – La Biennale di Venezia, titled *May You Live in Interesting Times*, curated by Ralph Rugoff.

Source: Courtesy the artist; *Aerocene* foundation; Andersen's, Copenhagen; Ruth Benzacar, Buenos Aires; Tanya Bonakdar Gallery, New York/Los Angeles; Pinksummer Contemporary Art, Genoa; Esther Schipper, Berlin. © Photography by Studio Tomás Saraceno, 2019.

In the concluding section I turn to an installation at the Venice Biennale to engage this question through a *counter-atlas* of clouds.

VI *Aero(s)cenes*

At the 2019 Venice Biennale Tomás Saraceno presented an installation at the Arsenale gallery. The installation was visible from stone pathways on the perimeter of one of the lagoons circling the gallery. A ramp enabled visitors to leave the stone path and step down into the installation titled: *Aero(s)cene: When breath becomes air, when atmospheres become the movement for a post fossil fuel era against carbon-capitalist clouds.* Two vitrines held artefacts from the *Aerocene* Community, including an *Aerocene* backpack. While standing next to the vitrines in October 2019, I looked out over the water at a delicately positioned sculpture of clustered dodecahedrons in the recognisable style of Studio Saraceno. Titled *On the Disappearance of Clouds* the sculpture represented an 'emergent cloudscape in a tidal scenography' that 'oscillate[d] throughout the day at the rhythm of the sea tides'.[72] The 'directors of this aerial theatre' were listed as 'planetary drifts of carbon-capitalist clouds [and] the anthropo-cumulus of the petrochemical pole of Porto Marghera towering over the bells of the San Francesco della Vigna Church'.[73] The text further elaborated: 'Clouds floating at the bottom of an ocean of air become the notations of a score'.[74] Indeed the installation was permeated by low and subtle hums: a sound work *Acqua Alta: En Clave de Sol* was transmitted by a series of six speakers positioned around the installation that acoustically amplified the ebb and flow of the Arsenale's canal waters on their six hour tidal cycle. As the sculptural cloud moved almost imperceptibly in the languid afternoon, I wondered: how does my awareness of this 'cloud' connect me to the tides, to carbon-capitalism or to the bells of a Venetian church? I struggled to untether my imagination from the spectacular installation, to feel the *pull* of an image, an allegory, an elsewhere.

While I contemplated this, I noticed that one of the vitrines held a document called *Emergent Cloudscapes* with a grid-like series of cloud photographs. I recognised the organisation of a typical Cloud Code Chart: a graphic tool circulated by National Weather Services for identifying clouds. In this exhibited version, however, the standard names and photographs of *cumulus* and *cirrus* were positioned alongside others, including the *demolition* cloud, *geoengineering* cloud, *atomic* cloud, *fog computing*, *mist computing*, the *asperitas* cloud and the *data* cloud. The introduction read:

> The present internationally adopted system of cloud classification was initiated in the nineteenth century, but since then, industrial activity has injected large amounts of carbon dioxide, aerosolar particulates and hot air into the atmosphere... New clouds are emerging, while others disappear, together with the nested ecosystems that they were regulating.[75]

This 'Cloudscapes' document employed the format of an elemental classification repopulated with cloud species that were anthropogenic, toxic, shapeless, algorithmic and mediated. In this way the document conveyed an eccentric 'everywhere' of clouds. It begged the question: *what is a cloud?*

In raising this question, the document referenced another artist-generated counter-atlas of the sky, Amy Balkin's *The Atmosphere: A Guide*, which similarly recalls the Cloud Code Chart in graphic form.[76] Instead of solidifying notions of air and the elements into the grid-like structure of a meteorological chart, Balkin's work demonstrates the reach of human gestures in an increasingly crowded yet still largely unfathomable atmosphere. The result is a critical assessment of the atmosphere as medium for military, geopolitical and scientific exploits. This practical and allegorical guide tests ways of relating the human to the atmospheric that also activates relations between bodies, elements, laws, war, weather and the planetarity of the climate.

Engaging with *Emergent Cloudscapes* and *The Atmosphere: A guide* suggests an answer to the question: *what is a cloud?* It can be solid, molecular, liquid, gaseous or algorithmic. It can be inside or outside the cell, the body, the house or the city. It can permeate interfaces, memories and archives. A cloud, according to these counter-atlases, is truly an 'aggregate, a nebulous set, a multiplicity'.[77] Impossible to enclose and contain, it creeps and haunts, and it is inconvenient and intrusive. In this way, *Emergent Cloudscapes* and *The Atmosphere: A Guide* give us a *nephotrope*: an ecology of cloud forms and figurations. This *nephotrope* transcends nature, culture and politics; it leaks beyond the individual cloud form, and it dramatically enlarges the possible relations between humans and elemental media.

Counter-atlases like these are significant not only because they parody existing classifications, but also as part of a larger effort to query how institutions, infrastructures and technologies orchestrate relationships between humans and elemental atmospheres. If our tools of orientation to sky and air emerged from Euro-American legacies of coloniality and naval domination, and if our greatest scientific institutions produce vital knowledge on climates and weather patterns that simultaneously enclose and disconnect local experience from planetary process, we require alternative rubrics, codes, vocabularies and imaginaries as necessary supplements, or as forms of refusal. As I have shown in this chapter and volume, art offers resources for querying power and politics in the air, precisely because existing frameworks depend so thoroughly on an inherited set of material imaginaries. As Whitehead teaches us, these imaginaries are as malleable as the worlds they engender.[78] Worlds can be lured elsewhere.

To trace elemental lures of cloud and sun is to enlarge possibilities for feeling the world otherwise. It is to question hierarchical and scalar models, to address the persistence of colonial-era imaginaries of the sky and to hold together potentially incongruous ideas and entities. It allows space for mesmeric apparitions. Yet, as DeLoughrey demonstrates, apprehending the uncanny lures of clouds and light involves locating another proposition

Figure 5.6 Tomás Saraceno, 2019; *Emergent Cloudscapes*, part of *Aero(s)cene: When breath becomes air, when atmospheres become the movement for a post fossil fuel era against carbon-capitalist clouds*, 2019; installation view of *On the Disappearance of Clouds*, 2019 at the 58th International Art Exhibition – La Biennale di Venezia, titled *May You Live in Interesting Times*, curated by Ralph Rugoff.

Source: Courtesy the artist; *Aerocene* Foundation; Andersen's, Copenhagen; Ruth Benzacar, Buenos Aires; Tanya Bonakdar Gallery, New York/Los Angeles; Pinksummer Contemporary Art, Genoa; Esther Schipper, Berlin. © Studio Tomás Saraceno, 2019.

in the form of an imperative. This is what Spivak calls the 'imperative to re-imagine the planet' where 'planetary imaginings locate the imperative in a galactic and para-galactic alterity' and 'cannot be reasoned into... self-interest'.[79] The planet, Spivak says, is 'a catachresis for inscribing collective responsibility as right. Its alterity, determining experience, is mysterious and discontinuous – an experience of the impossible'.[80] Grasping the uncanny otherness of light and the eerie multiplicity of clouds does not lead to a romantic elemental materialism. Rather this grasping or feeling enlarges the imperative to imagine a collective responsibility that is not an obligation but a right: a planetary lure for feeling.

Notes

1   For an account of this flight, see: Nicola Triscott, '*Aerocene* – Flight without borders', 17 November 2015. Available at: https://nicolatriscott.org/2015/11/17/aerocene-flight-without-borders/

2   A. Harris, *Weatherland: Writers & artists under English skies* (London: Thames & Hudson, 2015).

3   E. De Vet, *Weather-ways: Experiencing and responding to everyday weather* (Unpublished PhD thesis, University of Wollongong, Australia, 2014).

4   In making this statement I want to acknowledge and gesture to the fraught relationships of weather and climate, especially in times of climate crisis. The difficulty of claiming, based on scientific frameworks, that individual weather events are the result of climate change has resulted in myriad debates, controversies and confusions in the public sphere.

5   M. Hulme, 'Climate and its changes: A cultural appraisal', *Geo: Geography and Environment*, 2(1), 2015, p. 3.

6   N. Gaskill and A.J. Nocek (eds), *The lure of Whitehead* (Minneapolis, MN: University of Minnesota Press, 2014), p. 21.

7   Anton Mesmer was a German doctor of the early nineteenth century who is known for proposing a theory of a natural energy transference between animate and inanimate objects that he termed 'animal magnetism' and that later became known as 'mesmerism'. Isabelle Stengers (2015) describes the process through which Mesmer's technical practice or 'artifice' was denounced. This occurred through a commission appointed by King Louis the XVI (and including chemist Anton Lavoisier) that aimed not at determining whether Mesmer's treatment worked, but whether he had discovered a new physical fluid. By separating a judgement on the existence of Mesmer's fluid from the bodies and imaginations with whom Mesmer was working, the commission could conclude that there was no evidence for such a fluid, and therefore no evidence for the working of Mesmer's treatment. See: I. Stengers, *In catastrophic times: Resisting the coming barbarism* (London: Open Humanities Press, 2015). See also P. Adey, *Levitation: the science, myth and magic of suspension* (London: Reaktion, 2017) which explores many important links between practices of mesmerism and practices of levitation, for example in the nineteenth-century floating demonstrations of Alfred Sylvester.

8   For an elaboration of 'climatic-affective' atmospheres, see: B. Verlie, '"Climatic-affective atmospheres": A conceptual tool for affective scholarship in a changing climate', *Emotion, Space and Society*, 33, 2019, online publication ahead of print at: https://doi.org/10.1016/j.emospa.2019.100623

9   As elaborated on the *Fly with Aerocene Pacha* website:

On the 25th of January 2020, 'Aerocene Pacha' a fuel-free hot air balloon lifted a person into the sky, landing safely back on earth using only sun and air we all breathe, setting six unprecedented world records. We fly without fossil fuels, batteries, lithium, solar panels, helium, hydrogen and carbon emissions. This marks the most sustainable human flight in the history of aviation, being certified by the FAI (Fédération Aéronautique Internationale).

(*Fly with Aerocene Pacha*, 2020: np)

10 M. Serres, *Genesis* (Genevieve James and James Nielson, trans.) (Ann Arbor, MI: University of Michigan Press, 1995), p. 103.

11 Serres, *Genesis*, p. 103.

12 L. Daston, 'Cloud physiognomy', *Representations*, 135(1), 2016, pp. 45–71.

13 I. Spar, 'Mesopotamian creation myths', in: *Heilbrunn timeline of art history* (New York: The Metropolitan Museum of Art, 2000). Available at: www.metmuseum.org/toah/hd/epic/hd_epic.htm; see also: H. Torczyner, 'The firmament and the clouds, Rāqîaʻ and Shehāqîm', *Studia Theologica*, 1(1–2), 2008, pp. 188–196.

14 S.D. Gedzelman, 'Cloud classification before Luke Howard', *Bulletin of the American Meteorological Society*, 70(4), 1989, p. 382.

15 See: X. Qiu, *Treasury of Chinese love poems* (New York: Hippocrene Books, 2003), p. 133; *Zhinu* is also a genus of spiders in the family *Tetragnathidae*. Deities, cloud forms and weaving make interesting thematic entanglements.

16 See discussions of Ming Dynasty porcelain in: F. Brinkley, *China: Its history, arts and literature* (Vol. 1) (New York: J.B. Millet, 1902); C. Kelun, *Chinese porcelain: Art, elegance, and appreciation* (San Francisco, CA: Long River Press, 2004).

17 Ley dreamed of a systematised weather forecasting system that would employ telegraph networks and manuals of cloud, weather and wind in order to distribute 'sky-signs' to citizens (Ley, 1894). See: W.C. Ley, *Cloudland: A study on the structure and characters of clouds* (London: Edward Stanford Publishers, 1894).

18 J. Golinski, *British weather and the climate of enlightenment* (Chicago, IL: The University of Chicago Press, 2007).

19 Emphasis mine; Ruskin, cited in P.N. Edwards, 'A vast machine: Standards as social technology', *Science*, 304 (5672), 2004, p. 827.

20 L. Howard, *On the modifications of clouds* (3rd edn) (London: John Churchill and Sons, 1803).

21 Howard, *On the modifications of clouds*, p. 1.

22 Goethe, cited in Daston, 'Cloud physiognomy', p. 55.

23 Daston, 'Cloud physiognomy'.

24 WMO, *One hundred years of international co-operation in meteorology (1873–1973): A historical review* (Geneva: World Meteorological Organisation, 1973), p. 2.

25 WMO, *One hundred years*, p. 2.

26 Edwards, 'A vast machine'.

27 Daston, 'Cloud physiognomy', p. 57.

28 R. Scott, 'Hon. Ralph Abercromby', Obituary in *Nature*, 57, 1897, p. 55.

29 Daston, 'Cloud physiognomy', p. 67.

30 A. Mbembe and T. Nilsen, 'Thoughts on the planetary: An interview with Achille Mbembe', *New Frame*, 5 September 2019. Available at: www.newframe.com/thoughts-on-the-planetary-an-interview-with-achille-mbembe/

31 Daston, 'Cloud physiognomy', p. 59.

32 However, due to the efforts of Gavin Pretor-Pinney of the Cloud Appreciation Society, the Asperitas cloud was added to the International Cloud Atlas as a 'supplementary feature' (rather than a completely new type of cloud) in March 2017. For an account of the efforts to include the Asperitas cloud, see: J. Mooalem,

'The amateur cloud society that (sort of) rattled the scientific community', *The New York Times*, 4 May 2016. Available at: www.nytimes.com/2016/05/08/magazine/the-amateur-cloud-society-that-sort-of-rattled-the-scientific-community.html

33 Stengers, *In catastrophic times*, p. 124.

34 Daston, 'Cloud physiognomy', p. 49.

35 N. Wolchover, 'A world without clouds', *Quanta Magazine*, 25 February 2019. Available at: www.quantamagazine.org/cloud-loss-could-add-8-degrees-to-global-warming-20190225/

36 For a discussion on the limitations of standards see: P.N. Edwards, *A vast machine: Computer models, climate data, and the politics of global warming* (Cambridge, MA: MIT Press, 2010).

37 S. Engelmann and D.P. McCormack, 'Elemental aesthetics: On artistic experiments with solar energy', *Annals of the American Association of Geographers*, 108(1), 2018, p. 244.

38 F. Marion, *Wonderful balloon ascents: Or, the conquest of the skies* (London: Cassell, Peter and Galpin, 1874).

39 Marion, *Wonderful balloon ascents*.

40 This occurrence spurred Grenville Mellen to write, 'a woman in a balloon is either out of her element or too high in it' (Mellen, 1825: 185).

41 See: B. Fuller, *Critical path* (New York: St. Martin's Press, 1981).

42 W. McLean, 'Graham Stevens: atmospheric industries', *AA Files*, 70 (Winter), 2015, p. 140.

43 D. Brown, 'Sunstat: A balloon the rides on sunbeams', *Ballooning Magazine*, 11(2), 1978, p. 6.

44 As Elizabeth DeLoughrey writes, the celebratory low flying of vessels like the space shuttle *Endeavour* is provocative of visions of elsewhere; it, 'allows the spectator to participate in the making of an extraterritorial imagination' (DeLoughrey, 2014: 257).

45 I was invited to spend an afternoon talking to Julian Nott at his home in Santa Barbara, during which Nott spoke about the early experiments in solar ballooning and recounted his own experience to me. Sadly, Nott passed away in 2019 during a ballooning accident after a successful test flight of an experimental balloon over Warner Springs, CA. I am forever grateful for the time Nott gave to my questions and the contributions he made to aerostatic science and scholarship.

46 Triscott, '*Aerocene* – Flight without borders', np.

47 P. Chabard, 'Air crafted architecture', *Aerocene Newspaper* (Berlin: Studio Tomás Saraceno, 2015), p. 5.

48 K. Moe, 'The magnificence of *Aerocene*', *Aerocene Newspaper* (Berlin: Studio Tomás Saraceno, 2015), p. 14.

49 See: E. Chardronnet, 'COP21: Maiden flight of the zero carbon hot air balloon', *Makery*, 2015. Available at: www.makery.info/en/2015/11/30/cop21-le-premier-vol-de-montgolfiere-zero-carbone/

50 Tomás Saraceno and Dominic Michaelis have met and shared experiences in solar ballooning and ideas for membrane designs. These conversations had many direct influences on the design and construction of the *D-OAEC Aerocene*.

51 V. Agard-Jones, 'What the sands remember', *GLQ: A Journal of Lesbian and Gay Studies*, 18(2–3), 2012, pp. 325–346.

52 C. Sharpe, 'Antiblack weather vs. black microclimates', *The Funambulist*, 2017. Available at: https://thefunambulist.net/articles/32058

53 K. Hayashi, *From trinity to trinity* (Eiko Otake, trans.) (Barrytown, NY: Station Hill Press, 2010).

54 K. Barad, 'Troubling time/s and ecologies of nothingness: Re-turning, re-membering, and facing the incalculable', *New Formations*, 92(92), 2017, p. 63.

55 C.N. Waters, J.P. Syvitski, A. Gałuszka, G.J. Hancock, J. Zalasiewicz, A. Cear-reta and A. Barnosky, 'Can nuclear weapons fallout mark the beginning of the Anthropocene epoch?', *Bulletin of the Atomic Scientists*, 71(3), 2015, pp. 46–57.

56 Instead see: K. Yusoff, *A billion black Anthropocenes or none* (Minneapolis, MN: University of Minnesota Press, 2018).

57 E.M. DeLoughrey, *Allegories of the Anthropocene* (Durham, NC: Duke University Press, 2019).

58 Engelmann and McCormack, 'Elemental aesthetics', p. 245.

59 Brown, 'Sunstat'.

60 Agard-Jones, 'What the sands remember', p. 326.

61 For a discussion of 'responsible literacy' see: G.C. Spivak, *Death of a discipline* (New York: Columbia University Press, 2003); see also DeLoughrey's discussion of Spivak in: DeLoughrey, 'Planetarity', *Allegories of the Anthropocene*, pp. 63–97.

62 DeLoughrey, *Allegories of the Anthropocene*, p. 71; DeLoughrey's citations are taken from Grandy (2001: 77). Although I follow DeLoughrey in thinking with Spivak's 'planetarity' in relation to contemporary artistic performance and solar aesthetics, I am aware that Spivak's argument on planetarity was developed over many years through speeches and published writings, and was employed to attend to issues of migration in Europe as well as techniques of knowledge production in comparative literature. The capaciousness of planetarity enables DeLoughrey's arguments on nuclear culture and planetary 'wars of light' as well as, in my view, the arguments on elemental relations and imaginaries honed this chapter. See also: E. DeLoughrey, 'Radiation ecologies and the wars of light', *MFS Modern Fiction Studies*, 55(3), 2009, pp. 468–498.

63 'Becoming planetary' is explored through the thought of Gayatri Spivak and Silvia Wynter as a form of praxis in: J. Gabrys 'Becoming planetary', *e-flux Journal*, 2018. Available at: www.e-flux.com/architecture/accumulation/217051/becoming-planetary/

64 Grandy, cited in DeLoughrey, *Allegories of the Anthropocene*, p. 66.

65 This question has been at the core of the discussions of the Weather or Not Reading Group, a monthly London-based gathering for thinking and practicing on topics of weather, organised by Radio Earth Hold (Arjuna Neumann, Loure de Selys and Rachel Dedman), in which I participated from August 2019 to January 2020. At the time of writing, the Weather or Not Group includes: Jack Chrysalis, Andrea Zarza, Conor Lorigan, J.R. Carpenter, Jol Thoms, Rachel Dedman, Sanjita Majumder, Sophie Dyer, Tiago Patatas, James Goodwin, Anna Mikkola, Bianca Stoppani and Matt Earnshaw.

66 D. Hildyard, *The second body* (London: Fitzcarraldo Editions, 2017), p. 25.

67 Hildyard, *The second body*, p. 25.

68 Hildyard, *The second body*, p. 25.

69 For a discussion of the continuity and discontinuity of 'planetarity' see: Spivak, *Death of a discipline*.

70 Whitehead, *Process and reality*, p. 259.

71 Gaskill and Nocek, *The Lure of Whitehead*, p. 6.

72 T. Saraceno, '*Aero(s)cene*: When breath becomes air, when atmospheres become the movement for a post fossil fuel era against carbon-capitalist clouds' [exhibition text] Installation at the 58th International Art Exhibition – La Biennale di Venezia, *May you live in interesting times* (curated by Ralph Rugoff), 2019a, p. 1.

73 Saraceno, 'Aero(s)cene', p. 1.

74 Saraceno, 'Aero(s)cene', p. 1.

75 T. Saraceno, 'Emergent cloudscapes', in: *Aero(s)cene: When breath becomes air, when atmospheres become the movement for a post fossil fuel era against carbon-capitalist clouds.* Installation at the 58th International Art Exhibition – La Biennale di Venezia, *May you live in interesting times* (curated by Ralph Rugoff), 2019b, p. 1.

76 A. Balkin, *The atmosphere: A guide.* 2013/16. Available at: http://tomorrowmorning. net/atmosphere

77 Serres, *Genesis*, p. 103.

78 Trained as a mathematician, Whitehead was familiar with Albert Einstein's theories of relativity and often references these in his philosophical work. A lesser-known fact is that in 1922 Whitehead actually presented an alternative theory of gravitation that led to the same predictions as Einstein's field equations, although it was based on very different mathematical formulae. While Whitehead's 'non-covariant, action at a distance' model was largely ignored by cosmologists at the time, his proposal, also articulated in *Process and reality*, that our universe is ephemeral and might be replaced with another one with different laws of nature and different dimensions of the space-time continuum laid the ground for the theories of 'many universes' popularised in the 1980s. For Whitehead, nothing, not even the laws of physics, is immune to the possibility of change and reconfiguration.

79 G.C. Spivak, *An aesthetic education in the era of globalization* (Cambridge, MA: Harvard University Press, 2012), p. 340.

80 Spivak, *An aesthetic education*, p. 341.

References

Adey, P. (2017). *Levitation: The science, myth and magic of suspension.* London: Reaktion Books.

Aerocene (2020). *Fly with Aerocene Pacha.* Available at: https://pacha.aerocene.org/

Agard-Jones, V. (2012). What the sands remember. *GLQ: A Journal of Lesbian and Gay Studies*, 18(2–3), 325–346.

Balkin, A. (2013/16). *The atmosphere: A guide.* Available at: http://tomorrowmorning. net/atmosphere

Barad, K. (2017). Troubling time/s and ecologies of nothingness: Re-turning, re-membering, and facing the incalculable. *New Formations*, 92(92), 56–86.

Brinkley, F. (1902). *China: Its history, arts and literature* (Vol. 1). New York: J.B. Millet.

Brown, D. (1978) Sunstat: A balloon the rides on sunbeams, *Ballooning Magazine*, 11(2), 6.

Chabard, P. (2015). Air crafted architecture. *Aerocene Newspaper* (p. 5). Berlin: Studio Tomás Saraceno.

Chardronnet, E. (2015). COP21: Maiden flight of the zero carbon hot air balloon. *Makery.* Available at: www.makery.info/en/2015/11/30/cop21-le-premier-vol-de-montgolfiere-zero-carbone/

Daston, L. (2016). Cloud physiognomy. *Representations*, 135(1), 45–71.

DeLoughrey, E. (2009). Radiation ecologies and the wars of light. *MFS Modern Fiction Studies*, 55(3), 468–498.

DeLoughrey, E. (2014). Satellite planetarity and the ends of the Earth. *Public Culture*, 26(73), 257–280.

DeLoughrey, E.M. (2019). *Allegories of the Anthropocene.* Durham, NC: Duke University Press.

De Vet, E. (2014). *Weather-ways: Experiencing and responding to everyday weather.* Unpublished PhD thesis, University of Wollongong, Australia.

Edwards, P.N. (2004). A vast machine: Standards as social technology. *Science,* 304 (5672), 827–828.

Edwards, P.N. (2010). *A vast machine: Computer models, climate data, and the politics of global warming.* Cambridge, MA: MIT Press.

Engelmann, S., and McCormack, D. (2018). Elemental aesthetics: On artistic experiments with solar energy. *Annals of the American Association of Geographers,* 108(1), 241–259.

Fuller, B. (1981). *Critical path.* New York: St. Martin's Press.

Gabrys, J. (2018). Becoming planetary. *e-flux Journal.* Available at: www.e-flux. com/architecture/accumulation/217051/becoming-planetary/

Gedzelman, S.D. (1989). Cloud classification before Luke Howard. *Bulletin of the American Meteorological Society,* 70(4), 381–395.

Golinski, J. (2007). *British weather and the climate of enlightenment.* Chicago, IL: The University of Chicago Press.

Hayashi, K. (2010). *From trinity to trinity* (Eiko Otake, trans.). Barrytown, NY: Station Hill Press.

Harris, A. (2015). *Weatherland: Writers & artists under English skies.* London: Thames & Hudson.

Hildyard, D. (2017). *The second body.* London: Fitzcarraldo Editions.

Howard, L. (1803). *On the modifications of clouds* (3rd edn). London: John Churchill and Sons.

Hulme, M. (2015). Climate and its changes: A cultural appraisal. *Geo: Geography and Environment,* 2(1), 1–11.

Kelun, C. (2004). *Chinese porcelain: Art, elegance, and appreciation.* San Francisco, CA: Long River Press.

Ley, W.C. (1894). *Cloudland: A study on the structure and characters of clouds.* London: Edward Stanford Publishers.

Marion, F. (1874). *Wonderful balloon ascents: Or, the Conquest of the Skies.* London: Cassell, Peter and Galpin.

Mbembe, A., and Nilsen, T. (2019). Thoughts on the planetary: An interview with Achille Mbembe. *New Frame,* 5 September. Available at: www.newframe.com/ thoughts-on-the-planetary-an-interview-with-achille-mbembe/

McLean, W. (2015). Graham Stevens: Atmospheric industries, *AA Files,* 70 (Winter), pp. 138–143.

Mellen, G. (1825). *Sad tales and glad tales.* Boston, MA: S. G. Goodrich.

Moe, K. (2015). The magnificence of *Aerocene. Aerocene Newspaper* (p. 14). Berlin: Studio Tomás Saraceno.

Mooalem, J. (2016). 'The amateur cloud society that (sort of) rattled the scientific community', *The New York Times,* 4 May. Available at: www.nytimes. com/2016/05/08/magazine/the-amateur-cloud-society-that-sort-of-rattled-the-scientific-community.html

Qiu, X. (2003). *Treasury of Chinese love poems.* New York: Hippocrene Books.

Saraceno, T. (2019a). Exhibition text. In *Aero(s)cene: When breath becomes air, when atmospheres become the movement for a post fossil fuel era against carbon-capitalist clouds.* Installation at the 58th International Art Exhibition – La Biennale di Venezia, *May You Live in Interesting Times* (curated by Ralph Rugoff).

Saraceno, T. (2019b). Emergent cloudscapes. In *Aero(s)cene: When breath becomes air, when atmospheres become the movement for a post fossil fuel era against carbon-capitalist clouds.* Installation at the 58th International Art Exhibition – La Biennale di Venezia, *May You Live in Interesting Times* (curated by Ralph Rugoff).

Scott, R. (1897). Hon. Ralph Abercromby. Obituary in *Nature*, p. 55.

Serres, M. (1995). *Genesis.* (Genevieve James and James Nielson, trans.) Ann Arbor, MI: University of Michigan Press.

Sharpe, C. (2017). Antiblack weather vs. black microclimates. *The Funambulist.* Available at: https://thefunambulist.net/articles/32058

Spar, I. (2000). Mesopotamian creation myths. In *Heilbrunn timeline of art history.* New York: The Metropolitan Museum of Art. Available at: www.metmuseum. org/toah/hd/epic/hd_epic.htm

Spivak, G.C. (2003). *Death of a discipline.* New York: Columbia University Press.

Spivak, G.C. (2012). *An aesthetic education in the era of globalization.* Cambridge, MA: Harvard University Press.

Stengers, I. (2015). *In catastrophic times: Resisting the coming barbarism.* London: Open Humanities Press.

Torczyner, H. (2008). The firmament and the clouds, Rāqîaʻ and Shehāqîm. *Studia Theologica*, 1(1–2), 188–196.

Triscott, N. (2015). 'Aerocene – Flight without borders', 17 November. Available at: https://nicolatriscott.org/2015/11/17/aerocene-flight-without-borders/

Verlie, B. (2019). 'Climatic-affective atmospheres': A conceptual tool for affective scholarship in a changing climate. *Emotion, Space and Society*, 33, online publication ahead of print: https://doi.org/10.1016/j.emospa.2019.100623

Waters, C.N., Syvitski, J.P., Gałuszka, A., Hancock, G.J., Zalasiewicz, J., Cearreta, A., and Barnosky, A. (2015). Can nuclear weapons fallout mark the beginning of the Anthropocene Epoch? *Bulletin of the Atomic Scientists*, 71(3), 46–57.

WMO (1973). *One hundred years of international co-operation in meteorology (1873–1973): A historical review.* Geneva: World Meteorological Organisation.

Wolchover, N. (2019). A world without clouds. *Quanta Magazine*, 25 February. Available at: www.quantamagazine.org/cloud-loss-could-add-8-degrees-to-global-warming-20190225/

Yusoff, K. (2018). *A billion black Anthropocenes or none.* Minneapolis, MN: University of Minnesota Press.

6 Conclusion

New fault lines

I Book of clouds

This volume advances the notion of elemental lures for social scientific approaches to air and atmosphere by foregrounding the practices and experiments of the aerosolar arts. In conclusion, it is fitting to cite another parable of un-earthing. In Chloe Aridjis' novel *Book of Clouds*, set in early 2000s Berlin, a late August storm sweeps through the city over the weekend: 'the atmosphere was changing fast, the air driven by a new buoyancy'.[1] City dwellers hurry back to their homes. Aridjis' protagonist Tatiana can feel the change inside her Prenzlauer-Berg apartment building: 'the mounting pressure fought to enter, a tremendous suction at each point of entry, as if the harbinger winds were seeking refuge from an advancing sovereign'.[2] She struggles to close the windows: 'I could feel it trying to suck us into its mobile chaos'.[3] Once the storm passes and her building stops trembling, Tatiana walks through the rooms of her flat to see if anything has changed.

At first, the apartment looks unscathed. Then she notices 'row upon row of dirt, risen from the cracks between the floorboards'.[4] Each room is crossed by 'long caterpillars of dust'.[5] It is 'the dirt and dust of decades... drawn to the surface by the sheer force of suction'.[6] As she surveys the apartment, Tatiana has a 'growing hunch that although the storm had moved on, something in the building's very foundation had shifted, ever so slightly, revealing new fault lines'.[7]

In these scenes, Aridjis describes something remarkable: an elemental presence intrudes, quasi-invisibly, into Tatiana's apartment. This intrusion unsettles archives of dust, sedimentations of past inhabitants. The rune-like 'caterpillars' are histories unearthed: timescales unfold topologically on floorboards. Yet this parable is not just an account of the sublime power of a summer storm. It is an expression of the sensual and often uncanny propositions of atmospheric phenomena. Importantly, as conveyed by Aridjis, the storm does not elicit naïve wonder or awe. Rather, as ghosts levitate in grains of dust, new fault lines propagate through the envelope of the building. Whether they are visible or not, these fault lines trace elemental affinities and differences among those sheltering inside, as much from the storm as from the things it stirs up in the socio-political weather of Berlin.

The stories I have narrated in my own 'book of clouds' featured shared atmospheres, drifting elements and the more-than-meteorological weather. More specifically, I developed the notion of *elemental lures* with empirical attentions to three aerosolar initiatives: *Museo Aero Solar, Becoming Aerosolar* and *Aerocene*. I proposed that art amplifies the lures of aerial media, whether these are the shapes of clouds, uncanny rays of light, corridors of wind, or changes in the weather. At the same time, art invites us to sense at the limits of our perceptual capacities, to listen to traces of atmospheric currents and to imagine our bodies as clouds and as vessels of air. In shifting how we perceive, move and imagine, art enlarges aesthetic and political awareness of air and atmosphere.

Elemental lures take on many different shapes and qualities. In each of the preceding four chapters, I engaged different atmospheric sites, spaces and processes, marrying attentions to atmospheric affect with visceral awareness of the materials, institutions and processes hovering in the air. Through narrative-led writing, the voices of artists and collaborators were situated and central. In Chapter 2, I investigated the elemental lures of shared atmospheres through a focus on *Museo Aero Solar*. I argued that the collectively fabricated *Museo Aero Solar* sculpture creates infectious atmospheres that bind bodies to each other and to their elemental circumstances. By outlining part of *Museo Aero Solar's* history, I elaborated on the 'interstitial politics' of aerosolar fabrication, envelopment and flight.[8] Although they produce powerful 'spells', these interstitial atmospheres ask us to consider, within the multi-coloured envelope, the possibilities and limits of what is held in common. While sharing the same airspace, our bodies carry traces and marks related to our origins, manifesting the politics of location. As our vocal cords resonate frequencies in the same medium, our voices are implicated in economies of access and audibility. Thinking about the elemental lures of shared atmospheres enables an appreciation of being and breathing in common while recognising forms of power and politics in the air. These ideas informed the analysis of pedagogical atmospheres in Chapter 3, and the investigation of the politics of a post-launch studio gathering in Chapter 4. The notion of an elemental politics of location, adapted from feminist materialisms and cultural studies, echoed into themes of sand and soil memory in Chapter 5.

Bridging spaces inside and outside the aerosolar envelope, in Chapter 3, I considered the elemental lures of wind and weather. I discussed how perceptions of local winds and weather patterns in aerosolar practices open out onto perceptions of air-entry laws, insurance schemes, traffic control regulations and socio-political conditions. Furthermore, aerosolar sculptures intensify perceptions of the 'living air' and the 'weather-world' long before membranes are released; these *lures of perception* complicate linear accounts of processes of designing, making and launching. Turning to the complex relations between aerosolar envelopes, wind currents, weather conditions, technical devices and bodies, I used Whitehead's metaphysics to show how human perception emerges from more-than-human 'prehensions'. Moreover, the propositions of aerosolar sculptures, wind and weather phenomena and the pedagogical initiative *Becoming Aerosolar* helped me to outline a phenomenological attention to air that is not about distanced observation but is about 'interrogating' air for its simultaneous physical, affective and political qualities.[9] This phenomenological perspective has relevance for scholarly, artistic and activist dispositions to air and atmosphere; it also resonated in processes of tracking and tracing aerosolar entities in Chapter 4.

As APRS transmissions travelled through atmospheres, digital relays and mobile interfaces, two conjoined aerosolar sculptures drifted across the 'deserts' of the lower stratosphere, interrogating air for its creativity and novelty. Emphasising hybrid technological practices and atmospheres of suspension, Chapter 4 examined the elemental lures of transboundary air movements and regulated airspaces through the long-distance journey of the *Aerocene Gemini*. Observing the traces of the *Gemini* revealed that the sculptures ceased to be objects and were 'lost' to atmospheric currents: these, I suggested, are the *lures of movement* felt within the *Aerocene* Community during each aerosolar experiment. Still, we need to grasp that the moving traces and politics of these transnational flights are registered on human bodies. If we begin to see the practices of *Aerocene* as the traces or 'doings' of a political imaginary, we can also grasp the importance of a model of community governance that accounts for the weighing of political and practical work on different bodies. Elemental lures are thus deeply implicated in the politics of legibility, visibility and responsibility as well as in the movement of air masses and meteorological systems.

Each of the empirical chapters in this volume engaged with the relations between bodies and atmospheres and foregrounded the simultaneous meteorological and affective qualities of atmospheric spacetimes. In Chapter 5 I explored elemental lures of cloud and sun, and the interstices of weather and climate, through a focus on atmospheres of aerosolar levitation and the human-carrying *D-OAEC Aerocene* sculpture. I proposed that these choreographed levitations bring human bodies into proximity with meteorological entities. As such, I suggested that the imaginative lures of *Aerocene* can be complicated through an attention to the nebulous properties of clouds and the uncanny alterity of light. By reading the *D-OAEC Aerocene* launch at the White Sands National Monument as a *heliotrope,* we can better grasp its relationship to a historical lineage of artistic and solar experiments as well as to planetary ecologies of illumination and radiation.[10] Indeed, a heliotropic imaginary may inform many other instantiations of the aerosolar arts such as *Museo Aero Solar*, a sculpture and project built of materials originally derived from petrochemical by-products: chains of hydrocarbons fixed by ancient sunlight.

This book has traced elemental lures of wind and weather, sun and air through the experiments of the aerosolar arts. In conversation with social science literature on air and atmosphere, I have further established the place of art in theorising air and atmosphere and in untethering geographical material ontologies. Through a focus on lures, I illustrated how art moves, attracts and disrupts inherited forms of aerial knowledge and amplifies issues of power and difference. In the following section I consider how these insights inform wider projects of elemental aesthetics and politics. Then I make a call for the *elemental geohumanities*.

II Elemental aesthetics and politics

When you understand all about the sun and all about the atmosphere and all about the rotation of the earth, you may still miss the radiance of the sunset.

(Alfred North Whitehead[11])

The arguments, narratives and figurations in this book are informed by the non-anthropocentric philosophy of Alfred North Whitehead. The notion of elemental lures responds to Whitehead's proposals that, first, all entities and phenomena in the world exhibit 'emotional feeling',[12] second, the lures of the metaphysical world pre-exist our awareness of them, and third, the degree to which propositions, or lures, are 'relevant' or 'interesting' is related to their capacity to induce feeling, whether sensual, conceptual or imaginative. This has several consequences for aesthetics. In foregrounding elemental lures as independent from, yet apprehended and engaged by art, the arguments in this book sidestep what Gilles Deleuze calls the 'wrenching duality' of aesthetics, namely that it can refer to a 'theory of sensibility' or a 'theory of art'.[13] This duality manifests in recent attempts to grapple with the relations of the geo and the aesthetic, as signaled by scholars like Thomas Jellis.[14] Because of the ubiquity of propositions, elemental lures ripple across aesthetic categories and complicate discrete separations of 'art' and 'world'.

The primary referent of aesthetics in this volume is the *lure*. I introduced lures as feelings of push, pull and attraction. Lures, I suggested via Whitehead, are 'propositions' that elicit interest, induce differences and generate novelty. Yet lures are not tethered to human intellect, awareness or consciousness. Rather, they are registered by humans as part of the material and metaphysical luring of 'actual entities' or 'actual occasions'.[15] As demonstrated through the ecologies of *Aerocene* flights, for example, a handheld radio antenna is able to pick up APRS data because it 'prehends' bursts of noise on the radio spectrum at 144.8 MHz. We can think of these prehensions as 'propositional feelings' or lures in the sense that the antenna and receiver are 'pulled' into these sensory encounters by electromagnetic and elemental forces. Thus, in the Whiteheadian philosophy that has informed this volume, 'pulses of emotion' and feeling travel through materials, creatures and devices, participating in the blurring of ontological boundaries.[16]

To take this stance is to move closer to an understanding of aesthetics as *aesthesis*. It is to understand the world as 'a pure thinking-feeling behind which there is neither substance nor subject'.[17]

Furthermore, by addressing elemental aesthetics and politics through an approach and theory of lures, this book partially ungrounds the metaphysical referents (i.e. the human subject or the solid object) informing popular accounts of aesthetics and politics. As Mark Jackson argues, too often do geographers and others assume that emphasising aesthetics leads to expanded political capacities, or senses of the political.[18] Moreover, Jackson adds: 'If aesthetics is a way of rendering meaningful sense and sensibility for thought, then, importantly, we need to recognise that it also assumes, and mobilises, a particular idea of a thinking subject'.[19] Just as classical categories of the elements carry implicit assumptions about the metaphysics of thought that have fed into the projects and imaginaries of Empire, so too are western notions of aesthetics linked to Kantian theoretical frameworks and their limited (white, western, male) notions of the sensate subject.

Employing the lure as the referent of aesthetics complicates the human registers of judgement that inflect aesthetic theory. Beginning with lures rather than human *sensus communis* or 'power to judge' has enabled me to account for the ways that airy-elemental phenomena affect bodies and nonhuman entities alike. Attending to lures for feeling expands the potential registers of experience so that it becomes possible to suggest that two floating bodies, tethered together, prehend each other and their elemental circumstances in ways that open out onto concrete feelings of weather and climate. These propositions for feeling in the elements, or elemental lures, resonate with approaches to geo- and posthuman aesthetics that engage relational ontologies to trace aesthesis in nonhuman spheres.[20] My focus on elemental lures contributes to these debates by foregrounding an alternative perceptual referent that resists the duality of aesthetics, insists on the primacy of feeling and foregrounds a capacious elemental imaginary.

In our shared work on elemental aesthetics, Derek McCormack and I pose two questions about the relationship between the aesthetic and the elemental.[21] The first is the question of where and how the elemental shapes the conditions of aesthetic experience. This question highlights how different kinds of elemental media condition acts of sensing. Second, we examine 'the kinds of experiments, and experimental devices, that might contribute to enhancing capacities to sense the elemental in a range of circumstances'.[22] In this volume, I engage this second question by telling stories of artistic interventions into elemental spacetimes. The aerosolar arts offer 'experiments' and 'experimental devices' for making and staging shared atmospheres, amplifying perceptions of wind and weather, extending awareness into meteorological systems and manipulating imaginative tropes of air, light and heat. If earlier work on elemental aesthetics focused on 'becoming attuned to solar energy in new ways', the chapters in this volume attune to a wider range of atmospheric propositions, entities and ecologies.[23] At the risk of diluting the sense of air's 'elemental force' that comes with a focus on the relations of air and sun, this volume expands the horizon of elemental aesthetics by elaborating on the novelty and creativity of atmospheric propositions.

A key aim of the wider project of elemental aesthetics is to complement the representation of aerial entities with forms of non-representational awareness. This entails, for example, approaching the sun as 'pure process'[24] and approaching the clouds and winds as 'pure dynamism'.[25] While I gestured to the allegorical imagination in Chapter 5, I also engaged forces, textures and movements outside of representation. Moreover, through its larger commitment to tracking and tracing as both method and metaphor, this volume resonates with non-representational approaches insofar as 'a non-representational method involves an intensification of problems and requires staying with those problems for a while'.[26] In the accounts featured in these pages, 'staying with problems' is also 'to explicate the background of life and thought without presuming that the background is simply an inert 'context' or that the background is a mysterious, inaccessible, substance outside of all mediation'.[27] As I have shown, aerosolar sculptures are not objects drifting in neutral or empty air/space; they are informed and inflected by the substance of air and the many material and immaterial entities it carries and mediates. An aerosolar sculpture is 'lost' to its 'background' or collapses from its 'context' via the pressure of ambient air. This does not dissolve the relations at work or render them overly mysterious; rather, it amplifies aerial matters that were always already present when the sculptural 'object' seemed so unshakeably concrete.

How are elemental aesthetics and politics linked in this volume, and how do lures inform this relationship? If lures are not limited to human intellect or consciousness, their politics are not reducible to the expression and negotiation of human agency. In this way the politics of the lure has some affinities with the cosmopolitics or 'ecologised politics' developed originally by Isabelle Stengers and furthered by Bruno Latour, Steve Hinchliffe and Sarah Whatmore, among others.[28] This is a politics in which nonhuman entities and propositions are given the space to intervene in the stories we tell about them. It is a politics in which forms of 'know-how' and skill are given as much importance as intellectual creativity or ability.[29] I have addressed, for example, how the *D-OAEC Aerocene* sculpture intervenes in wider narratives of solar-aerostatic movement and acclimatised futures by reformatting human relationships to planetary process and difference. Just as the slow labour of connecting plastic interstices in *Museo Aero Solar* has relevance for the project's political momentum, so too do the plastic sheets themselves: they are stories that are woven and melted into the membrane.

However, more specifically, the politics of elemental lures is one in which feelings and sensibilities, whether fleeting or enduring, form the basis for micropolitical coalitions. Micropolitics is a Guattarian term that refers to the making of alternative dispositions and orientations, or 'mutant existential virtualities' towards present and future conditions.[30] Micropolitics describes the temporary cohesions of bodies, technologies and practices in an *Aerocene* flight, both in the sense that these cohesions generate alterative

dispositions to air and atmosphere and also in the sense that they intervene in the 'macropolitics' of aeromobility. Crucially, the relationships of
micropolitics to macropolitics are interstitial; to paraphrase Cindy Katz, as
contexts change, so too can designations of micro and macro or the boundaries between them.[31] Phrased differently, micropolitics is a reworking of
macropolitics from within because the two are intimately entwined.[32] A
micropolitical account is relevant because it alerts us, for example, to the
fraught relationships between the 'small stories' of the aerosolar arts and
the powerful actors, institutions and cultural programmes to which the aerosolar arts are tethered.

Guattari asserts that art generates possibilities for being and becoming in
the shadow of dominant material and economic assemblages.[33] Art and processes of subjectivisation access different 'speeds' and 'velocities' of feeling.[34]
Guattari calls these velocities 'ruptures': interventions in the fabric of worlds
in which 'something is detached and starts to work for itself'.[35] These interventions have relevance for resource ecologies, energy systems and elemental
attunements, since, heeding Isabelle Stengers' call, 'the earth [must be] taken
into account as a set of interdependent processes, capable of assemblages that
are very different from the ones on which we depend'.[36] One role of the arts,
Guattari and Simon O'Sullivan suggest, is to rupture from extractive and
fuel-burning resource systems and to access other models of energy infrastructure.[37] Indeed as Lucy Lippard notes in *The Lure of the Local*, such has
been the work of artists Helen Mayer Harrison and Newton Harrison, known
for their 'scientifically viable suggestions' about healing damaged landscapes
and resuscitating failed energy sources.[38] The Harrisons are known for their
proposals of 'models and metaphoric solutions' presented through stories.[39]
Although they are informed by enormous amounts of scientific research, it is
perhaps just as important that these propositions lure bodies and subjectivities beyond the petrochemical paradigm.

From the beginning of this volume, I engaged the somewhat paradoxi-
cal fact that bodies and entities are affected by a 'weather-and-water-world'
while absorbing specific elemental sites and histories via skin, blood and
bone. To elaborate on the politics of location and elemental air, I have taken
three interrelated approaches. First, I highlighted embodied experiences of
air, whether through stories of air pressing down on enveloped bodies or
through the body-becoming-cloud. These accounts are incompatible with
readings of the body as a discrete, subjective 'agent' or of the air as 'outside',
'object' or 'void'. Second, I foregrounded first-hand accounts in order to
situate myself and my own politics in these pages. Third, through narratives involving my colleagues in the aerosolar arts, I attempted to trace the
relationships between bodies and forms of power in the air. To this end, I
considered Débora Swistun's address inside *Museo Aero Solar* as a parable
for the uneven economies of shared atmospheres. I also unsettled interpretations of *D-OAEC Aerocene* launches by thinking through the historical
lineage of solar-aerostatic experiments and the elemental politics of irradiated landscapes. Elemental classifications and constructs became vehicles

for tracing imaginative investments of empire in the genesis of meteorology and elemental philosophy. Through these arguments, I articulated how aerial matters, processes and art forms are productive of alternative political coalitions while interacting with institutional atmospheres fraught with potential asymmetries.

In this volume, elemental lures propagate in, with and through air and atmosphere. However, ultimately, elemental lures are not attached to the aerosolar arts or, for that matter, to the air. In the next section, I trace elemental lures in water and wood, metal and soil to explore whether the lure has purchase for art and aesthetics in other elemental spheres. The rest of this conclusion gestures towards a wider programme of the *elemental geohumanities* to be expanded by other scholars and practitioners.

III Water and wood, metal and soil

How do lures operate in the non-aerial elements? Can we follow lures of the water and the earth like those of the air? How do artworks extend, amplify and express these lures? Detailed answers to these queries are beyond the scope of this volume, but here I will show how the lure may function as an aesthetic tool in non-aerial spheres. First, we might consider the elemental lures of water and arts that respond to and engage these lures. Like air, water is a fluid matter. It enters and leaves the body. It moves through the environment carrying many things with it. Water can push, pull and attract very much like air. The matters, metaphors and imaginaries of water have also elicited interest for un-mooring geographical ontologies.[40] A turn to the liquid and oceanic in cultural studies and the humanities has been called 'tidalectics', a term borrowed from the Barbadian poet and historian Kamau Brathwaite. Brathwaite describes tidalectics as 'the rejection of the notion of dialectic, which is three–the resolution in the third. Now I go for a concept I call 'tide-alectic' which is the ripple and the two tide movement'.[41] The concept of tidalectics is a lure for feeling in a Whiteheadian sense not only because it has elicited interest from both scholarly and artistic spheres, but also because it emerges from the characteristic movement of the tides. As Anna Reckin shows, this is a movement of the 'to and fro and back again'; it is registered on the pages of Brathwaite's *Barabajan Poems* via 'a tidalectic action' requiring nonlinear reading styles.[42] The 2017 exhibition *Tidalectics*, curated by Stefanie Hessler for Thyssen-Bornemisza Art Contemporary, explored tidalectics via the works of several artists 'whose practice is profoundly involved with the oceans'.[43] From Susanne M. Winterling's collaborations with dinoflagellate algae to Eduardo Navarro's *Hydrohexagrams*, an oceanic adaptation of the divinatory tool *I Ching*, these artistic practices centred on 'becoming more receptive to the liquid dimensions of our being'.[44] Although it is impossible to do justice here to the expanded lures of water or sea, the currency of Brathwaite's tidalectics in multiple spheres reveals that they are sensed and engaged by scholars and artists alike.

Whitehead offers, 'If we desire a record of uninterpreted experience, we must ask a stone to record its autobiography'.[45] His point is that human experience is always mediated by interpretation. Yet he also gestures to the idea at the core of his philosophy that all entities, even seemingly inert stones, feel and experience the world, albeit not at the same 'mental' level as humans.[46] Although they are less fluid than air or water, stone and earth act as lures in other ways that have also been apprehended by artists. For example, the Otolith Group's film *Medium Earth* narrates the tectonics of California's fault lines through interviews with individuals who have premonitions of earthquakes and other forms of seismic movement, locating perception as imbricated in continental pressures.[47] Without taking an anti-scientific stance, and echoing Whitehead, the film presents human interpretation of earthly events against the deep ripples of geological time. In this way *Medium Earth* not only responds to the elemental lures of tectonic movements but also presents entire sub-ecologies of citizen-led earth sensing. Through the resources of the lure, we can understand these ecologies of sensing through the 'pull' of awareness or the 'prehension' of propositions in vectors and fault lines. These lures are not only sensed by human bodies: *Medium Earth* visualises the expression of tectonic forces in boulder outcrops and hairline fractures, threading across dimly lit parking lots into Joshua Tree National Park. For these reasons, *Medium Earth* transcends its genre and becomes a matrix of elemental lures of geology and geomancy in the spatial geographies of the American Southwest.

As many have shown, the elemental is about more than defined substance or mediated immersion. It is also about the forces and state-changes that occur when different kinds of matter mix, amalgamate and become volatile to each other. What are the lures of elemental composition and combination? Yuriko Furuhata explains that the Eastern elemental philosophy of the *five phases* privileges mixtures and interactions, often following logics of generation and subjugation.[48] In Japan, Furuhata elaborates, practitioners of folk traditions attempt to calm strong winds by hanging iron sickles on tree branches, following the logic that 'metal conquers wood'.[49] These 'enchanted techniques of elemental control' belie 'a desire to manipulate the environment' that Furuhata traces into contemporary geoengineering and weather modification.[50] Thus, such practices illuminate the powerful imaginative lures in the seemingly 'unenchanted' tactics of cloud seeding.[51] In discussing these relations, Furuhata points to the prominence of alchemy in recent media theory. Although alchemy has not featured significantly in geographical literatures or in the untethering of geographical ontologies,[52] future work might explore the *allure* of the chemical elements via *alchemical* sensibilities and attentions.

Artists can aid geographers and others in these endeavours. In art, alchemical relations offer resources for probing the economies of elemental materials and questioning linear constructs of time. In her performance work, *A Transatlantic Periodic Table*, Ayesha Hameed addresses gold, bone, pearl, iron, copper and lead, citing their molecular compositions alongside their complicity in the economies of the Middle Passage. Chronicling her visit to a 'gold museum' closed decades ago, or feeling overwhelmed while holding a few glass beads discovered in the hold of a sunken slave ship, Hameed reminds us that these elements are haunted with colonial and racial struggles. Time, space and geography fold into strange transmutations as Hameed says towards the end of the performance: 'these past objects, séanced into the present, have undergone a *sea change*'.[53] If Furuhata suggests that geopolitical techniques of weather control are enchanted with the logic of elemental phases, Hameed proposes that the weather of colonial empire haunts life on the sea floor and on the surface. A series of elemental materials become lures for an alchemical inquiry into human suffering, colonial violence and the depths of the Atlantic.

The focus on ephemeral matters of air and atmosphere in this volume has not precluded an engagement with elemental histories and landscapes. In Chapter 5 I briefly considered the irradiated properties of the white sands of New Mexico in order to foreground elemental lures of radiation and solar light. It is worth further addressing lures of radioactive elements. Indeed, like an *Aerocene* sculpture, the trans-material and trans-corporeal flows of a radioactive element demand a commitment to tracking. Kate Brown's narration of the life and work of photographer Alexander Kupny is revealing of this. Working as a radiation monitor in Chernobyl from 1989, Kupny 'sought to grasp decaying radioactive isotopes as the elemental force burning at the center of the universe, the energy that gives life and can snatch it away'.[54] To this end, he entered the decaying sarcophagus of Reactor 4 wearing minimal protection and wielding a film camera. Kupny's eerie photographs are characterised by a snow-like texture, and the occasional 'comet' or 'spidery chandelier' of decaying isotopes.[55] Brown writes of these luminous phenomena: 'They are energy embodied. The specks are none other than cesium, plutonium, and uranium self-portraits'.[56] Feeling his way through labyrinths of irradiated matter, Kupny is lured by traces of elemental entities that remain at the edge of perceptibility yet further contaminate the cells of his body with each underground foray. Thinking with lures, we can understand Kupny's underground movements and his relentless search for isotopes as the *prehension* of the 'invisible map' of a nuclear event.[57] For this map to become legible, Kupny depends on the technical assemblage of a film camera that is lured to encounters with isotopes via cascades of radiation and the sensitivity of filmic emulsion.

An attention to radioactivity should hone our focus on other molecules, chemical units and mixtures that are toxic to diverse forms of life. If, as Denise Ferreira da Silva writes, the contemporary climate crisis is the material-energetic transformation of racial expropriation, do elemental lures offer ways of engaging these conditions? For Ferreira da Silva, the classical elements are useful 'metaphysical descriptors' for grasping the notion that colonial and racial violence is intrinsic to the accumulation of capital.[58] More specifically, however, it is the *heat* of the air, caused by the accumulation of greenhouse gases, that conveys relations of colonialism, capitalism and climate. Ferreira da Silva explains:

> The accumulation of atmospheric gases expresses (is equivalent to) the extent of expropriation and the intensity of the concentration of expro-priated internal (kinetic) energy of lands and labour facilitated by colo-niality and raciality.[59]

Such relations are not captured using linear constructs of time. They also work against separations of climate and culture, and unsettle metaphors of climate as a 'container' or 'stabiliser' that echo an intellectual history of enclosure and classification propagated by Euro-American ideologies.[60] The elements thus offer ways of 'interrupting' ideas of progress and articu-lating historical-climatic changes as *phase transitions*.[61] These are concepts and approaches that Ferreira da Silva explores with filmmaker Arjuna Neu-mann in their growing body of artistic work. Neumann and Ferreira da Sil-va's first collaborative film *Serpent Rain*, inspired by the story of a sunken slave ship, speaks 'from inside the cut between slavery and resource extrac-tion, between black lives matter and the matter of life, between the state changes of elements, timelessness and tarot'.[62] They pose the question: *what would become of the human if expressed by the elements?* Their second film *4 Waters: Deep Implicancy* follows routes through the Pacific, the Atlantic, the Indian Ocean and the Mediterranean Sea, conveying solid-fluid state changes and experimenting with time through the grammar of the film. In complement to Ferreira da Silva's philosophical thought, *Serpent Rain* and *4 Waters: Deep Implicancy* are audio-visual exercises in 'un-thinking the world' through the materiality and metaphysics of the elements.[63]

　　The lure offers important resources for engaging the propositions of the non-aerial elements. Elemental vectors of water and soil, and phase-shifts of isotope and alchemy produce lures: they push, pull, elicit interest and move humans, nonhumans and devices into novel orientations and configurations. Lures of the non-aerial elements are prominent in scholarly and artistic work, as the cur-rency of terms like tidalectics and the multi-media collaborations of Neumann and Ferreira da Silva make evident. Art apprehends the lures that manifest in tectonic folds and pressures, in the decay of subatomic particles and in sub-merged histories. Approaches to tracking and tracing are applicable in non-aerial spaces, even if they include very different 'maps' and methods. As in the air, one of the contributions of the lure in non-aerial spheres is that it places emphasis on prehension and process rather than on pre-existing relations, defi-nitions of subjectivity or categories of materiality and immateriality. A focus on the lures of the elements and on artistic practices that engage these lures might take shape in the geohumanities.

　　In referring to the geohumanities, I am aware of the varieties of practice, writing and thinking that have been identified with, or labelled under, this term.[64] The current momentum of the geohumanities has many influences: it is indebted to the rise of spatial thought in the social sciences and humanities since the 1980s, the growth of the digital humanities, the recent vogue around the prefix *Geo-* and the proliferation of the concept of the Anthropocene.[65] As Matthew Dear, Harriet Hawkins and Elizabeth Straughan have elaborated, the geohumanities may be reflected best by the swell of creative practice and collaboration undertaken by geographers and artists, fulfilling the call for such work made by Donald Meinig decades ago.[66] For this book, the geohumanities is neither new nor limited to geography. I understand it as an intensification of dialogue and shared methods between geography and the arts and humanities, contingent on the specific cultural, ecological and economic conditions of academic and artistic production today. In the following section I make a call for the *elemental geohumanities* and suggest what some of the concerns of this work might be.

IV Elemental geohumanities: writing, listening, experimenting

As signalled in my arguments about the elements, lures and art in this volume, the elemental geohumanities is already cohering. Elemental imaginaries, materials and concepts have motivated intellectual and creative work in geography. Geographers have also turned to the arts and humanities to witness, listen and experiment with the elements. Artists and humanities scholars have enlisted geographical thought in their engagements with elemental matters and media. Yet, this work is fragmented and has not been formally recognised. In the space left in this chapter, I outline some of this existing work and propose that the elemental geohumanities can further elemental geographies through the resources and critiques of art with particular attention to politics and difference.

As I outlined in this volume's introduction, the crafting of elemental attentions within cultural and creative geographies has a long history. Attentions to form, aesthetics, craft, media, design, narrative and poetics have long been part of geographical attentions to elemental substances and milieus. In this history and context, one of the contributions of art is to porously open geographical writing to the elements. As a recent example, we can turn to Kimberley Peters and Phil Steinberg's engagement with the ocean in their writing on 'wet ontologies'.[67] These authors employ extended excerpts of Yann Martel's *Life of Pi*, the poetry of William Wordsworth and Richard Collins' *The Land as Viewed from the Sea* to bring the wetness, fluidity, texture and materiality of the ocean into their prose. Giving space to the expressiveness of these artistic and literary works on the ocean is necessary for the project of un-mooring geographical and geopolitical ontologies.

Other geographers are more interested in how artworks make the elemental palpable or knowable. These scholars are perhaps less engaged in illustrating and evoking the elements than in investigating what elemental arts can do for geographical thought and practice. Derek McCormack's investment in the balloon is one example. In addition to the GHOST balloon campaign of the mid-twentieth century and the Google Loon project, in the book *Atmospheric Things* McCormack engages with many artistic sculptures, installations and figures, including the *Big Air Package* of Christo, the *Scattered Cloud* of William Forsythe, the inflatable interventions of Alfredo Jaar, the films of Andrei Tarkovsky and the large-scale installation *In Orbit* by Tomás Saraceno.[68] For McCormack, these artistic experiments offer, 'envelopes of captivation in which to experiment with thinking, feeling and moving differently'.[69] In a range of ways, they modify the form of the balloon, alter its aesthetic and political properties and demonstrate its value for concepts of affect, allure and the elemental.

Other works of geography craft responses to the question posed by Neumann and Ferreira da Silva: *what becomes of the human if expressed by the elements?*[70] In this body of work, art and aesthetics are understood as forms of sense making and feeling that implicate humans in elemental conditions. Drawing from a notion of art proposed by Elizabeth Grosz – namely, that art is a primary and 'vibrational' force among other energetic forces of Earth and cosmos – Katherine Yusoff explores paleolithic cave art to understand the ways human ancestors channelled and collaborated with 'inhuman' agencies.[71] Similarly, Nigel Clark investigates fiery and pyrotechnic arts as ancient world-building activities that are relevant for the thermal accelerations of the Anthropocene.[72] Kevin McHugh and Jennifer Kitson 'ignite the elemental' through experiments with an Infrared Camera in the deserts of the American Southwest.[73] David Paton narrates his practice as a sculptor of Cornish granite, arguing for an appreciation of matter as sentient.[74] Emerging from these works is not only the geopower of the elements, along with a sense of their force and anteriority.[75] Like the Otolith Group's film *Medium Earth*, these geographical contributions engage with art and aesthetics to recompose the figure of the human as thoroughly imbricated in the elements.

Among the creative proposals for sensing non-human, in-human or elemental worlds, methods of 'affinitive' or 'expanded' listening have become increasingly important. Michael Gallagher, Anja Kanngieser and Jonathan Prior write of

> the ways in which animals respond to sound; the electro-mechanical responses of listening technologies, from telephones to ultrasound scanners; or the ways in which seemingly inert materials are disposed to 'pick up' and respond to certain kinds of sonic vibration.[76]

In their emphasis on sensing beyond the concrete immediacy of the visual, these authors contribute to the larger project of querying metaphysical assumptions in geography. Listening to the elemental has been characterised in other ways too. Adey expands listening beyond the sonic when he calls for 'affinitive listening' (after Esther Leslie), asking geographers to attend to the chemical affects, repulsions and associations of the elements, especially the element of air.[77] If we think with notions of 'perturbation', as James

Ash proposes, we may become more observant of the listening practices of nonhuman objects and materials.[78] Thus, listening becomes a method, approach and ethic informed by creative registers that contribute to elemental geographies and geohumanities.

Just as geographers have engaged the arts and humanities for source material, sensory invitations and models of knowledge production, so too have humanities scholars and arts practitioners turned to geographical thought and practice. As many have attested, the arts and humanities' turn to geographical thought happened in part through a mutual interest in cartographic visualisations.[79] Furthermore, as artist-geographer Trevor Paglen asserts in an interview with Matthew Dear, the attraction of geography for artists may be due to the fact that geography offers 'a more robust theoretical framework' for understanding cultural economies and the production of space.[80] It is important, then, to also address the work of artists and humanities scholars who have specifically engaged with geography in aesthetic experiments with the elements.

V Elemental geohumanities: forensics, poetics, spectrality

Artists and humanities scholars, working in a variety of media, have turned to geography for tools to apply a 'forensic scrutiny' to worlds of matter.[81] As elaborated by Ursula Biemann in relation to her cinematic collaboration with Brazilian architect Paulo Tavares called *Forest Law*, an aesthetic practice that engages the scales of climate change and racial injustices must intervene in existing earth-narratives by focusing forensically on elemental matters: toxic drilling mud, copper, sulphur and the 'ur-liquids' of oil and water.[82] Biemann's film work undertakes a different kind of mapping, one that slows down in its elemental attentions, to '[let] imponderables come forth'.[83] Forensics is not the sole preserve of geography, however. We should remember the role of forensics as a state-driven power-science, and its mobilisation in architecture and law in the work of the research collective Forensic Architecture. Geographical forensics is visible in the disciplinary history of attentions to the production of space and power through material movements, from Ian Cook's paradigmatic 'follow the thing' exercise to the telling of 'geostories' with oil and ice.[84] For Nishat Awan, geographical forensics can be better attuned to the complexities of place, affect and subjectivity.[85] Thus, following Jo Sharp, Awan proposes that feminist geopolitics can inform forensic practices by foregrounding the economies of affect in acts of witnessing distant sites and struggles.[86]

These sensibilities are evident in artistic works that engage both forensics and the elements. In her body of audio-visual work bridging themes of social repression and hydropower projects in Columbia called *Dammed Landscapes*, Carolina Caycedo repurposes satellite imagery to produce multi-scaled documents of resistance to extraction, pollution and dispossession.[87] Caycedo thus subverts and annotates the instrumental use of digital and surveillance technologies to present the perspectives of local inhabitants that persist despite the footprint of hydropower. In her remarkable book *Drift*, Caroline Bergvall synthesises Old Norse, English and French with Forensic Architecture's research into the 'Left-to-die Boat', a vessel carrying migrants from Libya to Europe through one of the most heavily regulated maritime zones in the Mediterranean Sea.[88] In a work of feminist geographical forensics, Bergvall diffracts and subverts the emphasis on calculability and expert knowledge in Forensic Architecture's account of the boat. Instead, using poetic verse, cartography, textual glitches and personal history, Bergvall conjures the affective economies of drift as both ancient and contemporary condition.

Bergvall's drifts should remind us of a wider and historical body of work emerging out of the intersections of textual craft, geography and earthy attunement. The mid-century magazine *Landscape*, edited by J.B. Jackson, was a nurturing space for geopoetic and geoaesthetic attentions to quiet dwelling on the land. The magazine published non-academic and experimental pieces by many of the geographers whose work offered the foundation to the 'humanistic turn' of the 1960s and 1970s.[89] Eric Magrane cites the relationship between poet Charles Olson and geographer Carl Sauer, and especially Olson's advocacy for a 'kinetics of poetry' as an important 'touchstone' in the development of both eco- and geopoetics.[90] Sensibilities of kinetics and an earthy-elemental account of the body feed into Sarah de Leeuw's *Geographies of a Lover* and Ilana Halperin's *Learning to Read Rocks*.[91] In these poetic works earthen matters, topographies and sedimentations pass through skin, veins and vessels: 'the geography closest in'.[92] Magrane further makes a case for 'geopoetics as geophilosophy', a project that he argues, 'employs the widest conception of a poem, where climate change and the Anthropocene themselves—embodied in the form of a massive dam, for example —are large-scale geopoetics'.[93] In this expansive project, Magrane locates Yusoff's writing on geologic subjectivities and the earth-tracings of artists like Robert Smithson. Geopoetics as geophilosophy, then, is more about an ethics of 'radical experimentation' in shaping elemental media than it is about formal textual products.[94]

Other artists and humanities scholars have engaged the elements in a feminist mode through the relations between bodies and earthen matters. Artist Caitlin Berrigan and volcanologist Karen Holmberg, recipients of a GeoHumanities Creative Commission in 2018, travelled to Chaitén, Chile on an archaeological dig aiming to understand the cryptic hybrid *spider and vulva* inscriptions on the walls of a system of caves in this volcanic region. Berrigan produced a film based on this encounter as part of her evolving episodic narrative *Imaginary Explosions,* which follows a group of transfeminists who construct sub-surface networks and conspire with Earth's volcanos to erupt simultaneously.[95] In other works, the body moves underground without physically digging or descending beneath the surface. Flora Parrott employs digital printing, fabric and sculpture to invoke subterranean spaces and imaginaries related to her visit to the PETAR cave network in Brazil.[96] Libita Clayton's internationally recognised exhibition *Quantum Ghost* features an installation reminiscent of a mine shaft in which very low vibrations and ghostly voices render palpable a much larger subsurface of resource extraction, fragmented memory and black life.[97] In addition to creative experiments with earth and the underground, the aforementioned artworks are also exercises in spectral geography.

In briefly outlining a body of work in the *elemental geohumanities*, I do not mean to claim that the elemental is a new focus in geography, the arts and humanities, or to fall back on an uncritical account of nature via the elements. Rather, I am proposing that elemental imaginaries, materials and concepts have been significant in motivating intellectual and creative work in geography and the humanities, and have done so in particular ways. Geographers have turned to the arts and humanities to witness, listen and experiment with the elements. Artists and humanities scholars have enlisted or echoed geographical thought in their engagements with elemental matters and media. Through the resources of geography, poetics can encompass the shape of a large-scale dam, elemental phenomena can tell us of their role in historical power topologies, and forensic analysis can be situated and submerged. These contributions can and do productively inform geographical literature that queries the elemental as material ontology, as immersive milieu and as unit of molecular matter. However, there is scope for the elemental geohumanities to contribute more concretely to questions of power, politics and difference.

This work might take shape in several ways. First, geohumanities approaches to the elemental can further unsettle geography's material imaginary. In this volume, I showed that art has played an important role in untethering geographical ontologies of ground and solidity towards aërographies and aerial poetics. I also gestured to the work of geographers who engage with art to articulate 'wet' ontologies. However, one of the projects of the elemental geohumanities may be to further probe the material assumptions haunting geographical research. This is an ontological and epistemological as well as political project because, as noted earlier, the metaphysics of Anglo-American geography assumes certain elemental categories and human bodies as the reference points for scholarship. Thus, the elemental geohumanities can more forcibly critique the taken-for-granted material ontologies and metaphysics of geography and can do so through a capacious attention to elemental media, from chemical units, to alchemies and elemental phases. Second, the kinds of information allowed 'in' via artistic practices in the geohumanities can shift hierarchies of knowledge. This has particular relevance in relation to the elemental because the elements are not abstract categories: they are personal and political. The elemental geohumanities might therefore foreground accounts of the elements that demonstrate how they are held within the networks, tissues and organs of human and nonhuman bodies, and how they inform social and political worlds. In this way, and drawing from the contributions of feminist, antiracist, queer and indigenous scholars, the elemental geohumanities might further develop the elemental politics of location as a tactic, ethic and method. This would include furthering accounts of location and difference as intrinsic to elemental experience and illuminating the power matrices that inform and condition these experiences. In lieu of closure, I return to the elemental lures of air and atmosphere and the uncanny un-earthing of another August storm.

VI *Bura*

In Zaraće, on the island of Hvar, where I began this volume, August storms are well-known phenomena. One can predict that a storm is coming when the cumulus clouds dancing on the mountain ridges of the mainland become epic towers of cumulonimbus and tumble towards the island. The sea-swell can capsize a boat in the bay if tended by inexperienced hands. These storms are often electric: tendrils of lightning flash in between sky, cloud, rock and tree.

In the late summer of 2019, I spent a long night alone with my partner in the stone house my grandfather built in Zaraće in the 1960s. A storm hit the bay around 10pm. For several hours, it felt like the storm was talking to itself over the island. Lightning sent out its call. The answering waves of thunder rippled over the mountains and into the crevices of the jagged *Karst* rock formations by the sea. Inside the limestone walls of our house, which stands under an olive tree above the water, the shock waves registered in other ways too. Here, there were no wooden floorboards that could release the dust of the decades, but there were legions of elaborate moths, spiders, dragonflies and gecko-like *tantarelas* that sought refuge inside. Ladybugs climbed up the outer walls in multitudes to sneak through cracks in the window frames. In the morning we found that the iron-rich soil had washed up to and over the terrace, leaving bright red wavy lines like topographical maps.

The material relationships we witness in storms like this, and those observed by Aridjis in *Book of Clouds*, are elemental relationships.[98] The clouds release their burdens 'like the decanting of ten thousand aquariums', while the winds stir the sea into foam and spray.[99] Dirt is aerated, displaced or liquified. The elements, defined at the beginning of this volume as ontological categories of matter, environmental milieus or chemical units, are not adequate descriptors for the 'suction', the 'mobile chaos' or the 'new fault lines' expressed in these events. In other words, weathering a storm in Zaraće is the sensing of an elemental grammar unique to the bay and the island: something unspoken and inexpressible and yet intimately familiar to the moths, spiders and lizards. Hence, as I have suggested elsewhere, searching for a language adequate to the elements may be an impossible task.[100] Or it may be, as Whitehead teaches us, that our theories of the elements are not fixed truths but are propositions that emerge from the elements themselves. These propositions have greater or lesser relevance or 'interest' for each 'nexus of occasions', each storm. In turn, the storm expresses 'novelty' in its meetings with the land and the sea, and in the earthen, de-sedimented histories left in its wake.[101]

To weather a storm, and to take Whitehead to heart, is to be humbled by the force of the elements and to recognise the forms of sensing, signalling and luring that occur among and between elemental matters and media. This is not necessarily to invoke the sublime, but to grasp that there are material and metaphysical 'ripples' below and above human awareness that become palpable to us in their amplification. Particles of ancient dust are sucked from wooden crevices by a gale; the *Karst* rocks of Zaraće change colour, prehending the energy of the clouds above them. As I have proposed in this volume, art responds to and engages the elements, feeling-into the 'deserts' of the stratosphere or the atmosphere of a room. Thus, art's role in elemental philosophy and geography is far from tangential or illustrative. Instead, elemental scholarship requires the resources and gestures of art if it is to respond to the elements with scholarly precision and creative humility.

Figure 6.1 The author and her brother, Elliott, observing *Bura* from the rocks of Zaraće bay in 1995.

Source: Photograph by Steve Engelmann.

Notes

1 C. Aridjis, *Book of clouds* (New York: Black Cat, 2009), p. 13.
2 Aridjis, *Book of clouds*, p. 14.
3 Aridjis, *Book of clouds*, p. 14.
4 Aridjis, *Book of clouds*, p. 15.
5 Aridjis, *Book of clouds*, p. 15.
6 Aridjis, *Book of clouds*, p. 15.
7 Aridjis, *Book of clouds*, p. 15.
8 P. Pignarre and I. Stengers, *Capitalist sorcery: Breaking the spell* (A. Goffey, trans.) (Basingstoke: Palgrave Macmillan, 2011).
9 B. Anderson and J. Wylie, 'On geography and materiality', *Environment and planning A*, 41(2), 2009, pp. 318–335.
10 E.M. DeLoughrey, *Allegories of the Anthropocene* (Durham, NC: Duke University Press, 2019).
11 A.N. Whitehead, *Science and the modern world* (Cambridge, MA: Cambridge University Press, 2011 [1925]), p. 199.
12 A.N. Whitehead, *Process and reality*, corrected edition (New York: The Free Press, 1978 [1929]).
13 G. Deleuze, *The logic of sense* (M. Lester, trans.) (New York: Columbia University Press, 1990), p. 260.
14 See: T. Jellis, 'Geoaesthetics redux', *Social and Cultural Geography*, 17(8), 2016, pp. 1169–1172; H. Hawkins and E.G. Straughan (eds), *Geographical aesthetics: Imagining space, staging encounters* (Farnham: Ashgate, 2015).
15 Whitehead, *Process and reality*.
16 Whitehead, *Process and reality*, p. 163.
17 Massumi, cited in Jackson, 'Aesthetics, politics and attunement', p. 17.
18 See: M. Jackson, 'Aesthetics, politics, and attunement: On some questions brought by alterity and ontology', *GeoHumanities*, 2(1), 2016, pp. 8–23.
19 Jackson, 'Aesthetics, politics and attunement', p. 10.
20 See e.g. N. Clark, 'Sex, politics, and inhuman artistry', *Dialogues in Human Geography* 2(3), 2012, pp. 271–275; D. Dixon, H. Hawkins and E.G. Straughan, 'Of human birds and living rocks: Remaking aesthetics for post-human worlds', *Dialogues in Human Geography*, 2(3), 2012, pp. 249–270; K. Yusoff, 'Geologic subjects: Nonhuman origins, geomorphic aesthetics and the art of becoming in human', *Cultural Geographies*, 22(3), 2015, pp. 383–407.
21 S. Engelmann and D.P. McCormack, 'Elemental aesthetics: On artistic experiments with solar energy', *Annals of the American Association of Geographers*, 108(1), 2018, pp. 241–259.
22 Engelmann and McCormack, 'Elemental aesthetics', p. 244.
23 Engelmann and McCormack, 'Elemental aesthetics', p. 246.
24 See: Engelmann and McCormack, 'Elemental aesthetics'.
25 G. Bachelard, *Air and dreams: An essay on the imagination of movement* (E. Farrell and F. Farrell, trans.) (Dallas, TX: Dallas Institute Publications, Dallas Institute of Humanities and Culture, 1988).
26 B. Anderson and J. Ash, 'Atmospheric methods', in P. Vannini (ed.), *Nonrepresentational methodologies: Reenvisioning research* (Abingdon: Routledge, 2015), p. 48.
27 Anderson and Ash, 'Atmospheric methods', p. 48.
28 See Stengers' development of 'cosmopolitics', in: I. Stengers, 'Cosmopolitiques', *Natures Sciences Societes*, 5(3), 1997, p. 83; See also works that extend and develop cosmopolitics, including: B. Latour, *Pandora's hope: Essays on the reality of science studies* (Boston, MA: Harvard University Press, 1999); S. Hinchliffe, M.B. Kearnes, M. Degen and S. Whatmore, 'Urban wild things: A

cosmopolitical experiment', *Environment and Planning D: Society and Space*, 23(5), 2005, pp. 643–658; S. Whatmore and C. Landström, 'Flood apprentices: An exercise in making things public', *Economy and Society*, 40(4), 2011, pp. 582–610.

29 Hinchliffe et al., 'Urban wild things', p. 655.

30 F. Guattari, *Chaosmosis: An ethico-aesthetic paradigm* (P. Bains and J. Pefanis, trans.) (Sydney: Power Publications, 1995), p. 120.

31 C. Katz, 'Towards minor theory', *Environment and Planning D: Society and Space*, 14(4), 1996, pp. 487–499.

32 T. Jellis and J. Gerlach, 'Micropolitics and the minor', *Environment and Planning D: Society and Space*, 35(4), 2017, pp. 563–567.

33 Guattari, *Chaosmosis*.

34 Guattari, *Chaosmosis*.

35 Guattari, *Chaosmosis*, p. 132.

36 I. Stengers, *Thinking with Whitehead: A free and wild creation of concepts* (Cambridge, MA: Harvard University Press, 2011), p. 163.

37 See Guattari, *Chaosmosis*; see also: S. O'Sullivan, 'Guattari's aesthetic paradigm: From the folding of the finite/infinite relation to Schizoanalytic Meta-modelisation', *Deleuze Studies*, 4(2), 2010, pp. 256–286.

38 L. Lippard, *The lure of the local: Senses of place in a multicentered society* (New York: New Press, 1997), p. 185.

39 Lippard, *The lure of the local*, p. 185.

40 See for example: P. Steinberg and K. Peters, 'Wet ontologies, fluid spaces: Giving depth to volume through oceanic thinking', *Environment and Planning D: Society and Space*, 33(2), 2015, pp. 247–264; K. Peters and P. Steinberg, 'The ocean in excess: Towards a more-than-wet ontology', *Dialogues in Human Geography*, 9(3), 2019, pp. 293–307.

41 Braithwaite, cited in P. Naylor, *Poetic investigations: Singing the holes in history* (Evanston, IL: Northwestern University Press, 1999), p. 145.

42 A. Reckin, 'Tidalectic lectures: Kamau Brathwaite's prose/poetry as sound-space', *Anthurium: A Caribbean Studies Journal* 1(1), 2003, p. 3.

43 S. Hessler, *Tidalectics* [exhibition publication] (Vienna: Thyssen-Bornemisza Art Contemporary, 2017), p. 2.

44 Hessler, *Tidalectics*, p. 30.

45 Whitehead, *Process and reality*, p. 15.

46 Whitehead, *Process and reality*, p. 248.

47 Otolith Group (Kodwo Eshun and Anjalika Sagar), *Medium Earth*, 2013. Available at: http://otolithgroup.org/index.php?m=project&id=152

48 Y. Furuhata, 'Of dragons and geoengineering: Rethinking elemental media', *Media + Environment*, 1(1), 2019, np. Available at: https://mediaenviron.org/article/10797-of-dragons-and-geoengineering-rethinking-elemental-media

49 Furuhata, 'Of dragons and geoengineering', np.

50 Furuhata, 'Of dragons and geoengineering', np.

51 Furuhata, 'Of dragons and geoengineering', np.

52 Alchemy is mentioned or featured tangentially in the works of: A.M. Romero, J. Guthman, R. Galt, M. Huber, B. Mansfield and S. Sawyer, 'Chemical geographies', *GeoHumanities*, 3(1), 2017, pp. 158–177; P. Adey, 'Air's affinities: Geopolitics, chemical affect and the force of the elemental', *Dialogues in Human Geography*, 5(1), 2015, pp. 54–75.

53 A. Hameed, *A trans-Atlantic periodic table*. Performance during the Soot Breath Symposium, Goldsmiths University, London, 18 May 2019, np.

54 K. Brown, 'Marie Curie's fingerprint: Nuclear spelunking in the chernobyl zone', in A.L. Tsing, N. Bubandt, E. Gan and H.A. Swanson (eds), *Arts of living on a damaged planet: Ghosts and monsters of the Anthropocene* (Minneapolis, MN: University of Minnesota Press, 2017), p. G36.

55 Brown, 'Marie Curie's fingerprint', p. G41.

56 Brown, 'Marie Curie's fingerprint', p. G41.

57 R. Braidotti, *Nomadic subjects: Embodiment and sexual difference in contemporary feminist theory* (New York: Columbia University Press, 1994).

58 D. Ferreira da Silva, 'On heat', *Canadian Art*, 2018. Available at: https://canadianart.ca/features/on-heat/

59 Ferreira da Silva, 'On heat', np.

60 For a discussion of the relations of climate and culture, and the rationale for conceiving of climate as a 'container' or 'stabiliser' of weather, see: M. Hulme, 'Climate and its changes: A cultural appraisal', *Geo: Geography and Environment*, 2(1), 2015, pp. 1–11.

61 Ferreira da Silva, 'On heat', np.

62 A. Neumann and D. Ferreira da Silva, 'Serpent rain', interview with Margarida Mendes, *Vdrome*, 2016. Available at: www.vdrome.org/neuman-da-silva

63 Neumann and Ferreira da Silva, *Serpent Rain* [film].

64 Remembering the critiques of Braidotti (2019), I am also wary of the role played by the posthumanities within neoliberal tendencies in contemporary universities, at the same time as I want to celebrate the platforms that posthumanities research and scholarship offer.

65 T. Cresswell, 'Space, place, and the triumph of the humanities', *GeoHumanities*, 1(1), 2015, p. 4.

66 See for example: M. Dear, J. Ketchum, S. Luria and D. Richardson (eds), *GeoHumanities: Art, history, text at the edge of place* (Abingdon: Routledge, 2011); H. Hawkins, *For creative geographies: Geography, visual arts and the making of worlds* (Abingdon: Routledge, 2013); Hawkins and Straughan, *Geographical aesthetics*; For the call for an 'art' of geography see: D. Meinig, 'Geography as an art', *Transactions of the Institute of British Geographers*, 8, 1983, pp. 314–328.

67 Steinberg and Peters, 'Wet ontologies'; Peters and Steinberg, 'The ocean in excess'.

68 Although McCormack has written extensively on balloons and envelopes of other kinds, here I am thinking especially with: D.P. McCormack, *Atmospheric things: On the allure of elemental envelopment* (Durham, NC: Duke University Press, 2018).

69 McCormack, *Atmospheric things*, p. 76.

70 A. Neumann and D. Ferreira da Silva, *Serpent Rain* [film], 2016.

71 K. Yusoff, 'Geologic subjects: Nonhuman origins, geomorphic aesthetics and the art of becoming in human', *Cultural Geographies*, 22(3), 2015, pp. 383–407.

72 N. Clark, 'Fiery arts: Pyrotechnology and the political aesthetics of the Anthropocene', *GeoHumanities*, 1(2), 2015, pp. 266–284.

73 K. McHugh and J. Kitson, 'Thermal sensations – Burning the flesh of the world', *GeoHumanities*, 4(1), 2018, pp. 157–177.

74 D.A. Paton, 'The quarry as sculpture: The place of making', *Environment and Planning A*, 45(5), 2013, pp. 1070–1086.

75 Adey, 'Air's affinities'.

76 M. Gallagher, A. Kanngieser and J. Prior, 'Listening geographies: Landscape, affect and geotechnologies', *Progress in Human Geography*, 41(5), 2017, p. 622.

77 Adey, 'Air's affinities'.

78 For a discussion of 'perturbation' see: J. Ash, 'Rethinking affective atmospheres: Technology, perturbation and space times of the non-human', *Geoforum*, 49, 2013, pp. 20–28; For a discussion of 'more-than-human affinitive listening' using notions of perturbation, see: S. Engelmann, 'More-than-human affinitive listening', *Dialogues in Human Geography*, 5(1), 2015, pp. 76–79.

79 See for example: P. Crang, 'Cultural geography: After a fashion', *Cultural Geographies*, 17(2), 2010, pp. 191–201; H. Hawkins, L. Cabeen, F. Callard, N. Castree,

S. Daniels, D. DeLyser and P. Mitchell, 'What might GeoHumanities do? Possibilities, practices, publics, and politics', *GeoHumanities*, 1(2), 2015, pp. 211–232; L. Mogel, 'Disorientation guides: Cartography as artistic medium', in: M. Dear, J. Ketchum, S. Luria and D. Richardson (eds), *GeoHumanities: Art, history, text at the edge of place* (Abingdon: Routledge, 2011), pp. 187–195.

80 M. Dear, 'Experimental geography: An interview with Trevor Paglen, Oakland, CA, 17 February 2009', in: M. Dear, J. Ketchum, S. Luria and D. Richardson (eds), *GeoHumanities: Art, history, text at the edge of place* (Abingdon: Routledge, 2011), p. 24; a dedicated account of the intersections between geography and the arts, and an attempt to provide a series of guiding analytics for these intersections, can be found, from a geographical perspective, in: Hawkins (2013) *For creative geographies* and Dear et al. (2011) *GeoHumanities: Art, history, text at the edge of place*; and from a visual culture perspective, in Rogoff's *Terra infirma: Geography's visual culture* (2000).

81 U. Biemann, 'The cosmo-political forest: A theoretical and aesthetic discussion of the video forest law', *GeoHumanities*, 1(1), 2015, p. 163.

82 Biemann, 'The cosmo-political forest'. See also: U. Biemann, 'Deep weather', *GeoHumanities*, 2(2), 2016, p. 375.

83 Biemann, 'The cosmo-political forest', p. 167.

84 I. Cook, 'Follow the thing: Papaya', *Antipode*, 36(4), 2004, pp. 642–664; For 'geostories' see: R. Ghosn and E.H. Jazairy, *Geostories: Another architecture for the environment* (New York and Barcelona: Actar, 2019).

85 N. Awan, 'Digital narratives and witnessing: The ethics of engaging with places at a distance', *GeoHumanities*, 2(2), 2016, pp. 311–330.

86 See: J. Sharp, 'Reconsidering the material: A case for forensic geopolitics?', paper presented at the RGS-IBG Annual International Conference, University of Exeter, Exeter, UK, 2015. Available at: http://conference.rgs.org/AC2015/119

87 See: C. Caycedo, *Dammed landscapes*, 2013. Available at: http://carolinacaycedo.com/dammed-landscape-2013; for a longer discussion of Caycedo's *Dammed Landscapes* see: M. Gómez-Barris, *The extractive zone: Social ecologies and decolonial perspectives* (Durham, NC: Duke University Press, 2017).

88 C. Bergvall, *Drift* (Brooklyn, NY: Nightboat Books, 2014).

89 J.D. Blankenship, 'Midcentury geohumanities: JB Jackson and the "magazine of human geography"', *GeoHumanities*, 4(1), 2018, pp. 26–44.

90 E. Magrane, 'Situating geopoetics', *GeoHumanities*, 1(1), 2015, pp. 86–102.

91 S. deLeeuw, *Geographies of a lover* (Edmonton, AB: NeWest Press, 2012); I. Halperin, *Learning to read rocks*, performance at 'Feeling the Anthropocene: Air, Rock, Flesh' symposium, University of Edinburgh, UK, 2014. Available at: https://feelingtheanthropocene.wordpress.com/

92 A. Rich, 'Notes towards a politics of location', in: *Blood, bread and poetry* (New York: WW Norton and Company, 1986).

93 Magrane, 'Situating geopoetics', p. 94.

94 Magrane, 'Situating geopoetics', p. 94.

95 C. Berrigan, *Imaginary explosions* (Berlin: Broken Dimanche Press, 2018).

96 F. Parrott, *Fixed position* (2013–2014). Available at: www.floraparrott.com/new-page

97 L. Clayton, *Quantum ghost* [exhibition] (London: Gasworks Gallery, 2019).

98 J. Cohen, 'Elemental relations', *O-Zone: A Journal of Object-Oriented Studies*, 1(1), 2014, pp. 53–61.

99 Aridjis, *Book of clouds*, p. 14.

100 S. Engelmann, 'Adrift in the etheric ocean', *Dialogues in Human Geography*, 9(3), 2019, pp. 325–328.

101 For discussion of the 'nexus of occasions' and 'novelty' see: Whitehead, *Process and reality*.

References

Adey, P. (2015). Air's affinities: Geopolitics, chemical affect and the force of the elemental. *Dialogues in Human Geography*, 5(1), 54–75.

Anderson, B., and Wylie, J. (2009). On geography and materiality. *Environment and Planning A*, 41(2), 318–335.

Anderson, B., and Ash, J. (2015). Atmospheric methods. In P. Vannini (ed.), *Nonrepresentational methodologies: Reenvisioning research* (pp. 34–51). Abingdon: Routledge.

Aridjis, C. (2009). *Book of clouds*. New York: Black Cat.

Ash, J. (2013). Rethinking affective atmospheres: Technology, perturbation and space times of the non-human. *Geoforum*, 49, 20–28.

Awan, N. (2016). Digital narratives and witnessing: The ethics of engaging with places at a distance. *GeoHumanities*, 2(2), 311–330.

Bachelard, G. (1988 [1943]). *Air and dreams: An essay on the imagination of movement* (E. Farrell and F. Farrell, trans.). Dallas, TX: Dallas Institute Publications, Dallas Institute of Humanities and Culture.

Bergvall, C. (2014). *Drift*. Brooklyn, NY: Nightboat Books.

Berrigan, C. (2018). *Imaginary explosions*. Berlin: Broken Dimanche Press.

Biemann, U. (2015). The cosmo-political forest: A theoretical and aesthetic discussion of the video forest law. *GeoHumanities*, 1(1), 157–170.

Biemann, U. (2016). Deep weather. *GeoHumanities*, 2(2), 373–376.

Blankenship, J.D. (2018). Midcentury geohumanities: JB Jackson and the 'magazine of human geography'. *GeoHumanities*, 4(1), 26–44.

Braidotti, R. (1994). *Nomadic subjects: Embodiment and sexual difference in contemporary feminist theory*. New York: Columbia University Press.

Braidotti, R. (2019). *Posthuman knowledge*. London: Polity Press.

Brown, K. (2017). Marie Curie's fingerprint: Nuclear spelunking in the Chernobyl zone. In A.L. Tsing, N. Bubandt, E. Gan and H.A. Swanson (eds), *Arts of living on a damaged planet: Ghosts and monsters of the Anthropocene* (pp. 33–50). Minneapolis, MN: University of Minnesota Press.

Caycedo, C. (2013). *Dammed landscapes*. Available at: http://carolinacaycedo.com/dammed-landscape-2013

Clark, N. (2012). Sex, politics, and inhuman artistry. *Dialogues in Human Geography*, 2(3), 271–275.

Clark, N. (2015). Fiery arts: Pyrotechnology and the political aesthetics of the Anthropocene. *GeoHumanities*, 1(2), 266–284.

Clayton, L. (2019). *Quantum ghost* [exhibition]. London: Gasworks.

Crang, P. (2010). Cultural geography: After a fashion. *Cultural Geographies*, 17(2), 191–201.

Cresswell, T. (2015). Space, place, and the triumph of the humanities. *GeoHumanities*, 1(1), 4.

Cohen, J. (2014). Elemental relations. *O-Zone: A Journal of Object-Oriented Studies*. 1(1), 53–61.

Cook, I. (2004). Follow the thing: Papaya. *Antipode*, 36(4), 642–664.

Dear, M., Ketchum, J., Luria, S., and Richardson, D. (eds) (2011). *GeoHumanities: Art, history, text at the edge of place*. Abingdon: Routledge.

Dear, M. (2011). Experimental geography: An interview with Trevor Paglen, Oakland, CA, February 17, 2009. In M. Dear, J. Ketchum, S. Luria and D. Richardson (eds), *GeoHumanities: Art, history, text at the edge of place* (pp. 37–43). Abingdon: Routledge.

de Leeuw, S. (2012). *Geographies of a lover*. Edmonton, AB: NeWest Press.

Deleuze, G. (1990). *The logic of sense* (M. Lester, trans.). New York: Columbia University Press.

DeLoughrey, E.M. (2019). *Allegories of the Anthropocene*. Durham, NC: Duke University Press.

Dixon, D., Hawkins, H., and Straughan, E. (2012). Of human birds and living rocks: Remaking aesthetics for post-human worlds. *Dialogues in Human Geography*, 2(3), 249–270.

Engelmann, S. (2015). More-than-human affinitive listening. *Dialogues in Human Geography*, 5(1), 76–79.

Engelmann, S., and McCormack, D. (2018). Elemental aesthetics: On artistic experiments with solar energy. *Annals of the American Association of Geographers*, 108(1), 241–259.

Engelmann, S. (2019). Adrift in the etheric ocean. *Dialogues in Human Geography*, 9(3), 325–328.

Ferreira da Silva, D. (2018). On heat. *Canadian Art*. Available at: https://canadianart.ca/features/on-heat/

Furuhata, Y. (2019). Of dragons and geoengineering: Rethinking elemental media. *Media + Environment*, 1(1). Available at: https://mediaenviron.org/article/10797-of-dragons-and-geoengineering-rethinking-elemental-media

Gallagher, M., Kanngieser, A., and Prior, J. (2017). Listening geographies: Landscape, affect and geotechnologies. *Progress in Human Geography*, 41(5), 618–637.

Ghosn, R., and Jazairy, E.H. (2019). *Geostories: Another architecture for the environment*. New York and Barcelona: Actar.

Gómez-Barris, M. (2017). *The extractive zone: Social ecologies and decolonial perspectives*. Durham, NC: Duke University Press.

Guattari, F. (1995 [1992]). *Chaosmosis: An ethico-aesthetic paradigm* (P. Bains and J. Pefanis, trans.). Indiana, IN and Sydney: Power Publications.

Halperin, I. (2014). *Learning to read rocks*. Performance at 'Feeling the Anthropocene: Air, Rock, Flesh' Symposium, University of Edinburgh, UK. Available at: https://feelingtheanthropocene.wordpress.com/

Hameed, A. (2019). *A trans-Atlantic periodic table*. Performance during the Soot Breath Symposium, Goldsmiths University, 18 May.

Hawkins, H. (2013). *For creative geographies: Geography, visual arts and the making of worlds*. Abingdon: Routledge.

Hawkins, H., and Straughan, E.G. (eds) (2015). *Geographical aesthetics: Imagining space, staging encounters*. Farnham: Ashgate.

Hawkins, H., Cabeen, L., Callard, F., Castree, N., Daniels, S., DeLyser, D., and Mitchell, P. (2015). What might GeoHumanities do? Possibilities, practices, publics, and politics. *GeoHumanities*, 1(2), 211–232.

Hessler, S. (2017). *Tidalectics* [exhibition publication]. Vienna: Thyssen-Bornemisza Art Contemporary.

Hinchliffe, S., Kearnes, M.B., Degen, M., and Whatmore, S. (2005). Urban wild things: A cosmopolitical experiment. *Environment and Planning D: Society and Space*, 23(5), 643–658.

Hulme, M. (2015). Climate and its changes: A cultural appraisal. *Geo: Geography and Environment*, 2(1), 1–11.

Jackson, M. (2016). Aesthetics, politics, and attunement: On some questions brought by alterity and ontology. *GeoHumanities*, 2(1), 8–23.

Jellis, T. (2016). Geoaesthetics redux. *Social and Cultural Geography*, 17(8), 1169–1172.

Jellis, T., and Gerlach, J. (2017). Micropolitics and the minor. *Environment and Planning D: Society and Space*, 35(4), 563–567.

Katz, C. (1996). Towards minor theory. *Environment and Planning D: Society and Space*, 14(4), 487–499.

Latour, B. (1999). *Pandora's hope: Essays on the reality of science studies*. Boston, MA: Harvard University Press.

Lippard, L.R. (1997). *The lure of the local: Senses of place in a multicentered society*. New York: New Press.

Magrane, E. (2015). Situating geopoetics. *GeoHumanities*, 1(1), 86–102.

McCormack, D.P. (2018). *Atmospheric things: On the allure of elemental envelopment*. Durham, NC: Duke University Press.

McHugh, K., and Kitson, J. (2018). Thermal sensations—burning the flesh of the world. *GeoHumanities*, 4(1), 157–177.

Meinig, D. (1983). Geography as an art. *Transactions of the Institute of British Geographers*, 8, 314–328.

Mogel, L. (2011). Disorientation guides: Cartography as artistic medium. In M. Dear, J. Ketchum, S. Luria and D. Richardson (eds), *GeoHumanities: Art, history, text at the edge of place* (pp. 187–195). Abingdon: Routledge.

Naylor, P. (1999). *Poetic investigations: Singing the holes in history*. Evanston, IL: Northwestern University Press.

Neumann, A., and Ferreira da Silva, D. (2016a). *Serpent Rain* [film].

Neumann, A., and Ferreira da Silva, D. (2016b). *Serpent Rain*. Interview with Margarida Mendes, *Vdrome*. Available at: www.vdrome.org/neuman-da-silva

Neumann, A., and Ferreira da Silva, D. (2019). *4 waters: Deep implicancy* [film]. The Showroom Gallery, London.

O'Sullivan, S. (2010). Guattari's aesthetic paradigm: From the folding of the finite/infinite relation to Schizoanalytic Metamodelisation. *Deleuze Studies*, 4(2), 256–286.

Otolith Group (Kodwo Eshun and Anjalika Sagar) (2013). *Medium Earth*. Available at: http://otolithgroup.org/index.php?m=project&id=152

Parrott, F. (2013–2014). *Fixed position*. Available at: www.floraparrott.com/new-page

Paton, D.A. (2013). The quarry as sculpture: The place of making. *Environment and Planning A*, 45(5), 1070–1086.

Peters, K., and Steinberg, P. (2019). The ocean in excess: Towards a more-than-wet ontology. *Dialogues in Human Geography*, 9(3), 293–307.

Pignarre, P., and Stengers, I. (2011). *Capitalist sorcery: Breaking the spell*. (A. Goffey, trans.). Basingstoke: Palgrave Macmillan.

Reckin, A. (2003). Tidalectic lectures: Kamau Brathwaite's prose/poetry as sound-space. *Anthurium: A Caribbean Studies Journal*, 1(1), 1–16.

Rich, A. (1986). Notes towards a politics of location. In *Blood, bread and poetry*. New York: WW Norton and Company.

Rogoff, I. (2000). *Terra infirma: Geography's visual culture*. Abingdon: Routledge.

Romero, A.M., Guthman, J., Galt, R.E., Huber, M., Mansfield, B., and Sawyer, S. (2017). Chemical geographies. *GeoHumanities*, 3(1), 158–177.

Sharp, J. (2015). Reconsidering the material: A case for forensic geopolitics? Paper presented at the RGS-IBG Annual International Conference, University of Exeter, Exeter, UK. Available at: http://conference.rgs.org/AC2015/119

Stengers, I. (1997). Cosmopolitiques. *Natures Sciences Societes*, 5(3), 83–83.

Stengers, I. (2011). *Thinking with Whitehead: A free and wild creation of concepts.* Cambridge, MA: Harvard University Press.

Steinberg, P., and Peters, K. (2015). Wet ontologies, fluid spaces: Giving depth to volume through oceanic thinking. *Environment and Planning D: Society and Space*, 33(2), 247–264.

Whatmore, S.J., and Landström, C. (2011). Flood apprentices: An exercise in making things public. *Economy and Society*, 40(4), 582–610.

Whitehead, A.F. (1978[1929]). *Process and reality*, corrected edition. New York: The Free Press.

Whitehead, A.N. (2011 [1925]). *Science and the modern world.* Cambridge, MA: Cambridge University Press.

Yusoff, K. (2015). Geologic subjects: Nonhuman origins, geomorphic aesthetics and the art of becoming in human. *Cultural Geographies*, 22(3), 383–407.

Index

Note: *Italic* page numbers refer to figures and page numbers followed by 'n' denote endnotes.